THE COMPLETE GUIDE TO

EXTERNAL WALL INSULATION

By

Christopher J Pearson

First Published - September 2006
*Supplemented with CD Rom for Additional
Detailed Drawings & Standard Report Forms*

Published by
Wellgarth Publishing Limited
©C J Pearson 2006

Website - www.externalwallinsulation.com

THE COMPLETE GUIDE TO

EXTERNAL WALL INSULATION

By

Christopher J Pearson

THE COMPLETE GUIDE TO
EXTERNAL WALL INSULATION

Copyright © C J Pearson 2006

Conditions of Sale

Disclaimer

Distribution

York Publishing Services Limited & Wellgarth Publishing Limited
via www.externalwallinsulation.com

I S B N No   **0-9553636-0-8**
             978-0-9553636-0-3

Printed by

York Publishing Services Limited
64 Hallfield Road
Layerthorpe
York YO31 7ZQ

www.yps-publishing.co.uk

Contents

CD ROM

Supplement providing detailed drawings and
Standard Forms in Pdf Format
(Requires Adobe Acrobat Reader)

Acknowledgments

This manual is essentially based on the considerable information provided by the many specialist material suppliers, Trade Associations, Professional Bodies and individuals.

I confirm that I have made every effort to trace copyright holders where necessary and would welcome any information in tracing copyright holders who have not been contacted or acknowledged.

Many thanks to the following Authorities and Companies who have supplied data, information or general assistance in the creation of this manual:-

BRE Certification Limited   (BBA)
British Board of Agrement   (WIMLAS)
Insulated Render &Cladding Association   (INCA)
Insulated Concrete Formwork Association   (ICFA)
British Standards Institute.   (BSI)
BRC Special Products Limited
Kilwaughter Chemical Co Limited
Renderplas Limited
Eurobrick Limited
WEMICO Limited
VWS Technologie AM BAU
A.P.G.Management   (Klimex)
Carrs Paints Limited
Ejot Limited
Fischer Fixings Limited
Expamet Building Products
Easipoint Marketing Limited
Telling (Architectural Terracotta) Ltd
Telling (Lime Products) Ltd
Wetherby Systems Ltd
Wall-Reform Limited

Additional thanks go to Roy Pickavance, BSc, FRICS, FCIOB, FCIArb, and John Scanlon for their invaluable help and assistance in submitting comments, making suggestions and proof reading the text.

Introduction

The author, Christopher J Pearson, a qualified Building Surveyor, whilst formally the Managing Director of Epsicon Limited, successfully externally insulated many thousands of buildings of all descriptions over the years up to the year 2000, since retiring, the Company has progressed under new management.

A former Council Member of the External Wall Insulation Association, now the Insulated Render & Cladding Association, and as a former member of the original Technical Sub-Committee, Christopher Pearson was instrumental, with others, in the composition and production of the original technical data sheets for the Association.

This guide reviews and analyses the techniques currently available to insulate externally the walls of all types of buildings. Since the major energy saving interest in the United Kingdom is concerned with up-grading dwellings and public sector buildings, such as schools and hospitals, those are the areas on which attention will be focussed. However, some consideration has been devoted to the use of external insulation on the walls of new buildings and where air-conditioning is used.

The past decade has seen increasing energy costs. Both at home and in the workplace where we have become used to expecting a high standard of thermal comfort. Real incomes have declined, energy costs increased, so the effect of costs and conditions has produced the demand for insulation. The Building Industry is now refurbishing more and more houses due to lack of new house building and the effect of increased planning restrictions. But what of the many thousands of occupied houses which are uninsulated or poorly insulated? The only true remedy avoiding massive upheavals is to externally insulate them, giving them a new lease of life both with comfort and appearance.

Air-conditioning costs are considerably higher to provide than heating, external wall insulation works equally as well to save costs to cool buildings as well as heat them. Institutional and public buildings which are air-conditioned would benefit greatly by the application of external wall insulation in respect of their energy running costs.

Specifiers and installers who are just beginning to look closely at external insulation systems are well advised to make a detailed examination of what is on offer and what constitutes an effective, strong, long lasting and maintenance free insulation system.

Consideration of non-cavity wall insulation systems will include both internal and external insulation. Internal systems have a number of disadvantages insomuch that dwellings need to be evacuated, dimensions become critical and the internal life is disrupted. Whereas the external option allows uninterrupted occupation of the building and an efficient contract.

Technical considerations leave the internal option still with "cold-bridges" and doubts about interstitial condensation. Whereas external insulation can successfully avoid the latter and create a thermal store of the structure which improves the whole concept of energy use and control comfort levels.

There are many millions of homes in the United Kingdom today now insulated with some form of cavity wall insulation. There are also an equal number of domestic buildings which are of solid wall construction and are not suitable for cavity wall insulation together with many thousands of houses with reduced cavities, classed as "hard to heat" houses and not capable of receiving any form of approved cavity wall insulation. To all of these properties some form of exterior or interior insulation may be the only solution.

Historically, energy supplies have been under strain from the mid 70s when oil became a political weapon by middle east countries. Prior to this period, there was little or no consideration to the use of fossil fuels, coal was plentiful and energy conservation was not on any agenda.

Some fuel related workers had free fuel as part of their pay, transport in particular used fuel very inefficiently and domestic central heating was just beginning to be promoted in substantial numbers. All of these factors led to substantial increases in the use of fossil fuels, environmental pollution, particularly with "acid rain" and the increase of CO^2 levels resulting in the beginning of global warming.

There is the modern dilemma of CO^2 production being "invisible" as against the 19th century burning of fossil fuels in open fires causing acid rain and smog. Whilst the invisible gases of CO^2 are now resulting in global warming, the original types of pollution resulted in poor health and the erosion of buildings. All as a result of ignorance of the effects of excessive use of energy and its inefficient method of burning.

The first energy crisis in 1973 promoted thoughts as to energy conservation and future policies. Building regulations began to be improved and "U" values became part of the vocabulary. Prior to the energy crisis, there was little or no emphasis on the environmental aspects of burning fossil fuels and their emissions and certainly no thought given to the possibility of energy running-out.

There was little or no awareness of these problems and in some areas a complete lack of technical knowledge in how to avoid such problems in the future. In the late 1980s there was the landmark discovery of the "hole in the ozone layer" which prompted much awareness on energy transmissions.

During the early 1980s, external wall insulation began to appear in the UK, whilst being available for many years earlier on the European continent. After a slow start, several systems became available and certifying and testing by the British Board of Agrément was born.

The use of external, rather than internal insulation offers a number of practical advantages:-

1.   There is little or no disturbance to the occupants and the possible additional cost of temporary re-housing is avoided.

2.   The problems of insulating internally around or behind internal fittings eg. in kitchens and bathrooms, but also

cupboards, shelves, power points etc are avoided.

3.   Potential cold bridges at all internal/external wall and ceiling junctions are reduced or eliminated.

4.   In a complex building, it is comparatively easier to ensure that the whole of the external surface is insulated.

5.   The whole wall structure is protected so external surfaces are prevented from receiving further deterioration by being exposed to the weather.

There are many thousands of high energy consuming public sector and commercial buildings, both of a traditional and system built construction where the application of exterior wall insulation may be desirable if the thermal performance of the building is to be brought up to the current standards.

If external wall insulation is installed the running costs of air-conditioning may become an economic possibility.

Fixing a thermal insulation layer to the exterior of a building is an extremely potent method of generally up-grading the wall's thermal performance and the building's habitability.   Almost invariably internal condensation problems are alleviated. As a result of the increased thermal stability, comfort levels are improved. The external appearance is improved markedly and whole estates of sub-standard dwellings can undergo a remarkable face-lift with all its consequent social and political benefits. Finally, the whole building obtains an additional structural protection which may significantly increase its expected life.

For external wall insulation to be successful a number of physical criteria have to be satisfied if the main object of heat retention is to be effective.

Insulation materials lose their heat retaining properties if they become moist. Resistance to rainwater penetration and reduction of the risk of interstitial condensation are also important physical criteria.

In order to satisfy the former condition the surface should be waterproof and crack free, since cracks of above 0.5mm will admit some moisture.

Weatherproofing should not be achieved with a vapour impermeable layer since the building must still continue to breathe in the traditional manner. Although water is unlikely to penetrate the insulation from inside the building, water vapour may do so. This is a potentially unwelcome phenomenon as it is difficult to alter the position of the dew point in an externally insulated wall. This invariably occurs at the interfaces of differing materials.

By definition insulation is usually a fairly soft material and generally speaking the more open it is the better its insulation properties. A good quality insulation needs a tough skin to protect it against the usual casual impacts and even malicious damage or vandalism. This requires that both the reinforcing layer and fixing systems be robust and sufficiently strong to deter damage and carry the weight of the protective skin.

To appeal to the maximum number of Clients, Designers and Specifiers, any system offered should be capable of accepting a variety of finishes. That is not as easy as it seems since putting (say) roughcast on top of insulation is not the same as applying it onto the surface of ordinary brick or block.

Insulation immediately behind rendered finishes causes surface temperatures and other stresses to build up inside the whole system and to a much greater extent than if it were able to make use of the heat sink available in any other masonry substrate. The ability of masonry walls to retain heat also provides frost protection to the uninsulated external finish. Where external insulation is applied, this heat sink protection is eliminated and the finish has to be of sufficient quality to withstand the rigours of thermal shock and water penetration.

All of these properties suggests and implies that certification and testing is an essential requirement to any system.

Additional advantages and considerations in the use of external

5

wall insulation includes the following :-

1. "Incidental" reduction in transmitted sound into the building.

2. Availability of financial Government and Local Authority grants via energy providers

3. Carbon Dioxide emission reduction.

4. Reduced use of fossil fuels.

5. Low taxes (VAT) on energy.

The CD ROM Supplement provides
additional drawings and standard report forms
to be read in conjunction with the following parts.

PART ONE - Overview of External Wall Insulation

1.1   Why External Wall Insulation?

Government scientists have confirmed that climate change, caused by the increased concentration of green house gases, (CO_2 emissions), has caused increased earth temperatures. As Governmental pressure mounts to reduce this global warming, attention should be focussed on those areas most responsible for CO_2 emissions and action taken to make a real impact in reducing these. These areas include the reduced burning of fossil fuels for domestic heating, transport and industrial processes.

Traditional constructions have largely revolved around conventional bricks and blocks but during the last fifty years pre-cast concrete, timber frame and steel frames structures have been increasingly used. Building Regulations have largely governed the thermal performances of buildings and with the lack of appreciation of any need of energy conservation, these regulations were much inferior to those on the European continent until recent times.

External wall insulation is a particularly useful insulating method developed for exterior masonry walls over the last 30 years. First used in Europe for building refurbishment, recent innovations are now making the systems useful in new-build constructions providing all of the latest thermal requirements and still providing the traditional finishes that Architects and Designers require.

Traditional housing over the past 100 years usually comprised solid walls of masonry, either brick, blocks, concrete or stone. The method of heating provided was the provision of open fires resulting in plentiful ventilation and good comfort levels to habitable rooms. With the introduction of double glazing, sealed windows and the elimination of open fireplaces, internal wall condensation became a nightmare problem, particularly for public sector social housing.

The introduction of central heating systems substantially increased the total energy requirement, thus increasing the CO_2

emissions to unacceptable levels. It has become a requirement, therefore, to try to reduce or curtail these emissions at the same time control the consequential problems such as condensation.

Before & after photographs by courtesy of Wetherby Systems

Insulation in general and external wall insulation in particular can play a major role in energy saving strategy and specifiers also now accept that they have to be more environmentally aware.

This awareness of the technology and availability has led to the insulation industry experiencing a marked expansion in activity, particularly external wall insulation, both in private and public sectors.

Whilst all building types including Schools, Hospitals, Institutions and Public Buildings in general would benefit from external wall insulation, the principal use whereby the majority of walls are to be treated are in housing. This guide therefore focuses on the treatment of housing in general albeit the detailing and technologies are the same for whatever building is to be insulated with such methods.

1.2 House Types

House construction during the twentieth century, within the UK, is interesting and varied as the development of mass-produced industrialised housing evolved during the 1940s and 50s helping alleviate acute housing problems caused by two world wars together with erratic economic growth.

Traditional methods were extensively used in the construction of the housing stock, and industrialised methods were experimented with to speed construction. During this period, very little, if any, attention to the provision of insulation was given as energy was cheap and plentiful in the form of solid fuel. Basic building materials were in short supply and building regulations did not call for any particular standard of insulation either for walls or roofs. Industrialised methods proliferated, without regulation or control resulting in the present substandard aging housing stock.

The provision of open fire-places to burn solid fuel was the accepted method of heating which resulted in high levels of air-flow or air changes. All windows were single glazed and constructed in timber or steel. These windows were designed without any regard to draught proofing particularly with badly fitting opening-lights and no sealing strips. Whilst industrialized buildings were designed to be constructed quickly and efficiently, little was done to improve the design of windows or doors together with the thermal performance of external walls.

Building Regulations were dependant on Local By-Laws which were very basic in requirement and enforcement through local Authorities dependent on mainly unqualified inspectors. No consideration was given to the thermal performance of dwellings until circa 1965, even these requirements were for the provision of poor levels of insulation within roof spaces, no provision was given to walls until the late 1960s.

The following brief description of many of the house-types illustrates some of the many and varied construction types.

Traditional solid brick-block/Cavity Wall

Traditional constructions within the UK are due mainly to the easily accessible and plentiful brick and block material eminently suitable for the construction of housing and associated building. Over the years solid brick of 215mm (9") thickness has given way to the cavity wall, first constructed of two layers of brickwork, later the inner leaf being substituted for concrete blocks due to cost savings.

The insulation values currently in-force at this time were of no serious consideration with the primary concern being the prevention of ingress of water with the provision of a cavity. Combination walls comprising a cavity wall ground floor with a solid first floor was common practice prior to World War Two. "U" values were probably little known at this time but values of $1.6W/m^2DegC$ were probably achieved compared with the current requirement of $0.3W/m^2DegC$.

Typical Thermal Graphs

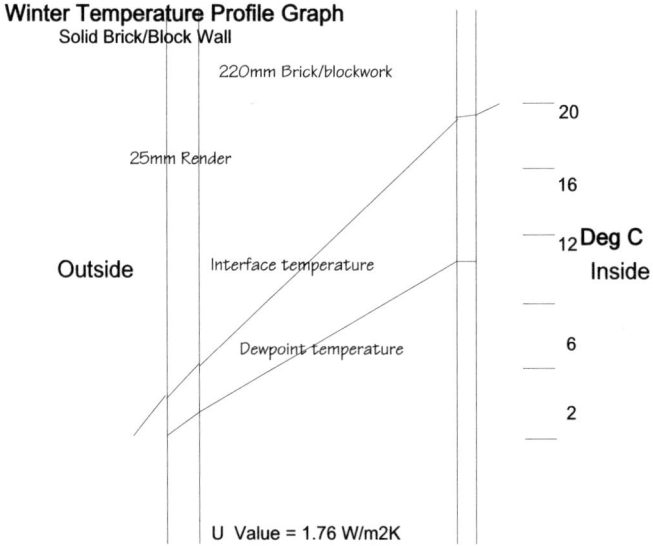

Winter Temperature Profile Graph
Solid Brick/Block Wall

220mm Brick/blockwork

25mm Render

Outside

Interface temperature

Dewpoint temperature

20

16

12 **Deg C**
Inside

6

2

U Value = 1.76 W/m2K

SECTION THROUGH BRICK/BLOCK WALL -UNINSULATED

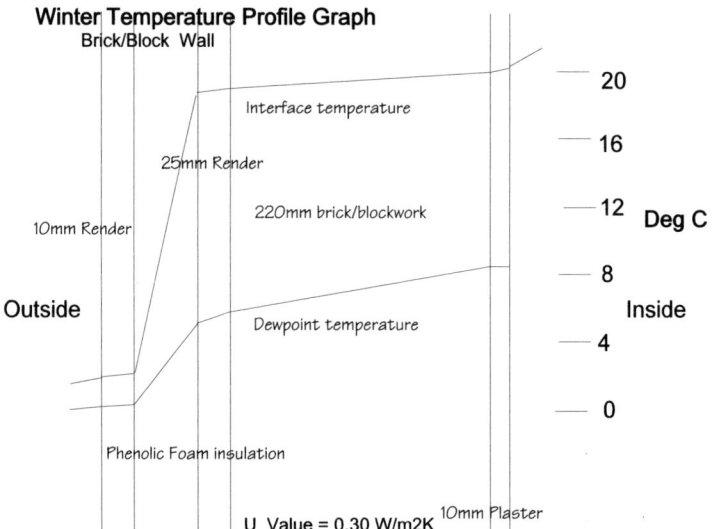

Winter Temperature Profile Graph
Brick/Block Wall

Interface temperature

25mm Render

10mm Render

220mm brick/blockwork

Outside

Dewpoint temperature

Inside

Phenolic Foam insulation

U Value = 0.30 W/m2K   10mm Plaster

20

16

12 **Deg C**

8

4

0

**SECTION THROUGH BRICK/BLOCK WALL WITH ADDED
EXTERNAL WALL INSULATION**

No-Fines Constructed Houses & Flats

A major programme of construction using no-fines concrete technology was instigated after WW2. This form of construction was fast and economical using locally sourced materials. Essentially "no-fines" is poured concrete without the small aggregate fillers, this enabled rapid construction methods due to the ability of the poured concrete to cure rapidly within the shutters, allowing quick assembly and stripping.

No-Fines concrete was made from locally extracted aggregates which could have been crushed limestone, rounded Thames gravel or flint gravel. All these gave different characteristics due to void formation, hardness of aggregate and distribution of cement paste.

Externally the system was rendered with a sand/cement render and internally insulation was occasionally installed but generally the internal surfaces were dry-lined with plasterboard.

Over the years this form of construction became very unpopular with tenants due to the uncomfortable and high-heating cost of the buildings. The lack of usage of the poorly designed heating systems by the tenants resulted in severe problems of internal wall surface condensation. Floor condensation was also prevalent due to the massive cold-bridges at all floor and party wall locations. High heating costs were usually totally unacceptable as tenants were usually those of low income groups.

Typical Thermal Graphs

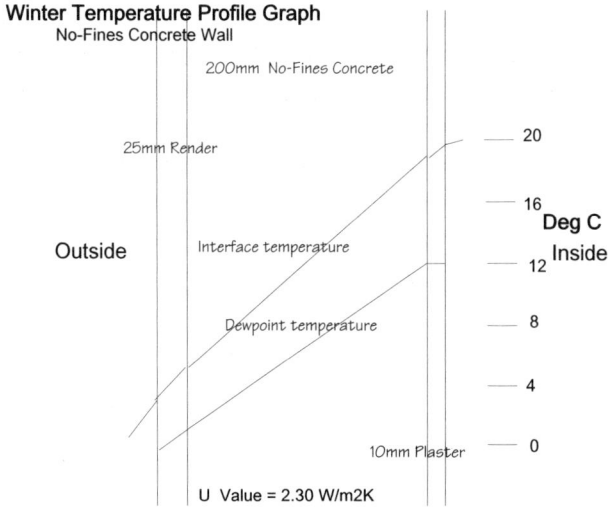

Winter Temperature Profile Graph
No-Fines Concrete Wall

200mm No-Fines Concrete

25mm Render

20

16

Deg C

Outside

Interface temperature

12

Inside

Dewpoint temperature

8

4

10mm Plaster

0

U Value = 2.30 W/m2K

SECTION THROUGH NO-FINES WALL -UNINSULATED

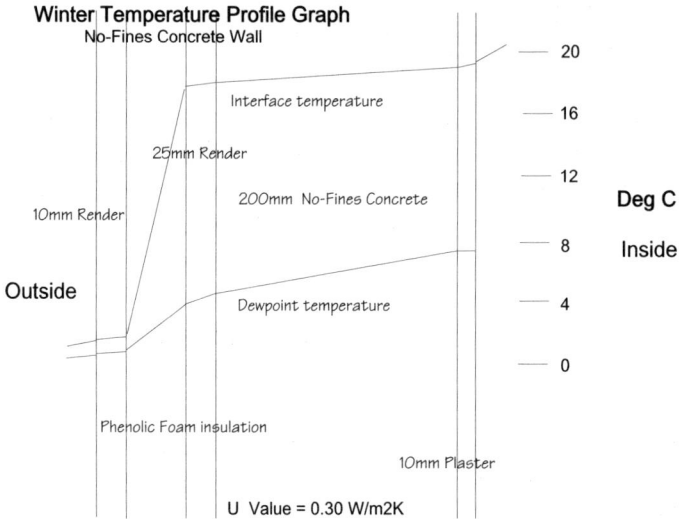

Winter Temperature Profile Graph
No-Fines Concrete Wall

20

Interface temperature

16

25mm Render

200mm No-Fines Concrete

12

10mm Render

Deg C

8

Inside

Outside

Dewpoint temperature

4

0

Phenolic Foam insulation

10mm Plaster

U Value = 0.30 W/m2K

SECTION THROUGH NO-FINES WALL WITH ADDED
EXTERNAL WALL INSULATION

14

Pre-Cast Concrete Houses

During the pre & post war period prefabrication of the new housing stock was undertaken to house the population quickly and economically.

This need was met with the introduction of many house types based on the use of pre-cast concrete for fast-track assembly. Insulation values were not a requirement at this time due to plentiful home produced energy sources and the lack of understanding of CO_2 emissions.

Several of these house types are listed elsewhere which hopefully form a typical selection of houses erected at this time, the average "U" values of walls is probably approximately $1.6W/m^2DegC$. Not only is this insulation value very low by today's standards, the extensive "cold-bridging" caused by the use of pre-cast concrete panels and columns is now totally unacceptable. From the thermal graph over-leaf, it can be seen where the intersection of the graph occurs, this is when condensation can form.

15

Typical Thermal Graphs

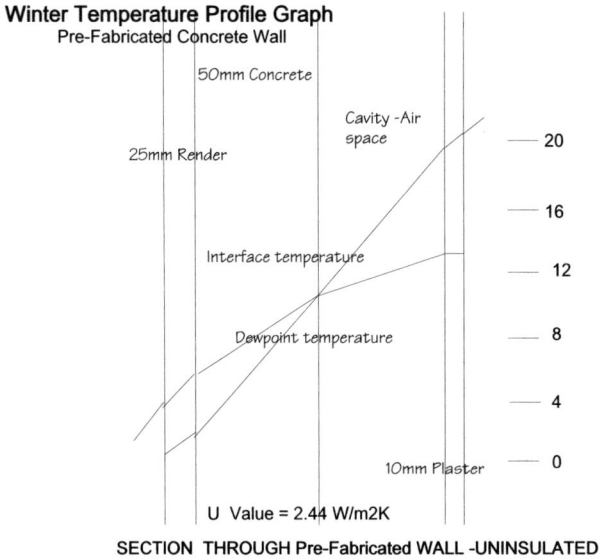

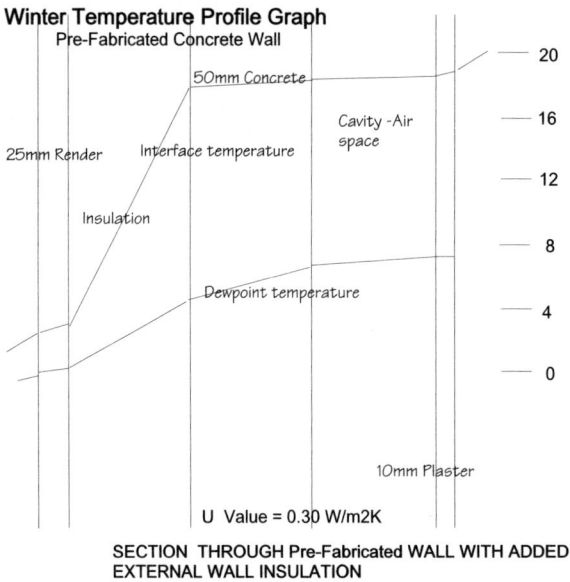

Winter Temperature Profile Graph
Pre-Fabricated Concrete Wall

50mm Concrete

Cavity -Air space

25mm Render

— 20

— 16

Interface temperature

— 12

Dewpoint temperature

— 8

— 4

— 0

10mm Plaster

U Value = 2.44 W/m2K

SECTION THROUGH Pre-Fabricated WALL -UNINSULATED

Winter Temperature Profile Graph
Pre-Fabricated Concrete Wall

— 20

50mm Concrete

Cavity -Air space

— 16

25mm Render   Interface temperature

— 12

Insulation

— 8

Dewpoint temperature

— 4

— 0

10mm Plaster

U Value = 0.30 W/m2K

SECTION THROUGH Pre-Fabricated WALL WITH ADDED
EXTERNAL WALL INSULATION

16

House Types in General

A selection of pre-cast Concrete-framed houses are described as follows:-

"Parkinson Framed Houses" are constructed using precast concrete columns and ring-beams in filled with cavity block work, without cavity ties. The first floor is constructed in timber.

"Woolaway Houses" are constructed of pre-cast concrete columns with pre-cast concrete in fill panels with block-work constructed gables.

"Underdown and Winget Houses" are constructed using pre-cast concrete wall slabs connected at vertical joints with column and ring-beams, the first floor being constructed in timber.

"Unity Houses" are constructed with storey height pre-cast concrete columns with unreinforced concrete cladding providing lateral stiffness of the structure.

"Wates Houses" are constructed with pre-cast reinforced concrete load-bearing panels, externally finished and structurally jointed.

"Cornish House Types" are constructed with external concrete columns with pre-cast in fill slabs, mansard roofs and erected in a wide variety of sizes, storeys and forms.

"Reema Houses" are constructed similar to Wates Houses in self-finished storey height pre-cast concrete structural panels.

"Smith Houses" are constructed using large pre-cast lightweight aggregate concrete panels usually with a built-in brick slip facade.

Many other house types were erected at this time including BISF Houses, Howard Houses which were steel framed and also "Calder Houses" which were timber framed.

Obtain from BRE - full publication list of PRC houses and their details.

Buildings Unsuitable for External Wall Insulation

Period or Medieval Buildings Medieval churches, barns, Country Houses and Listed Buildings of all types may prove difficult or even undesirable to clad with a modern form of insulation.

Planning constraints by numerous Local Authorities, National Parks, Land Owners and others may contribute to the external insulation constraints.

Buildings in Conservation Areas Conservation areas, many of which are in city centres, or towns of specific interest contain many examples of buildings with architectural interest. These buildings are usually well preserved and form a unique environment in the particular locality which will usually have architectural features which are difficult, if not impossible to replicate.

To clad these buildings with a conventional insulating cladding system would help destroy the architectural features and significantly alter the preserved environment.

Unstable buildings

Buildings considered unsuitable for external wall insulation include those constructed over old mine workings, subject to movement and subsidence together with those within flood plains and subject to the possibility of flooding.

Buildings that are unstable due to structural inadequacies must be structurally repaired prior to any external wall insulation cladding installation.

Agricultural Buildings

Most agricultural buildings are clad with vertical sheeting of metal or cementitious board construction.

Agricultural buildings are insulated only when required by the dictated use to the interior, if extra insulation is required additional internal insulation is the usual method adopted by farmers.

1.3 Principals of External Wall Insulation

The principals of wall construction are relatively simple and are as follows:-

1. Walls of any building serve to:-

    a, Keep out wind and rain.

    b. Preserve habitable living areas and privacy.

    c. Preserve the internal comfort levels of heat and humidity.

    d. Be structurally sound to support other elements essential for the building to function as designed.

2. Walls forming habitable areas must be capable of resisting :-

    a. Heat loss and/or gain.

    b. Water penetration.

    c. Formation of internal wall surface condensation.

    d. Formation of interstitial wall condensation.

    e. Excessive sound transmission.

Old type solid walls which had little or no insulation soon attracted water vapour to the internal face of the exterior walls following the introduction of a heat source, resulting in surface condensation. Internal surface condensation or moisture persisting on this wall face will eventually create the conditions to form mould growth or fungus giving rise to unhealthy living conditions leading to breathing problems.

The most effective method to correct this environment is to raise the internal wall surfaces so that the temperature at which water vapour will condense out of the internal air, is above the critical "dew-point" level. Frequent internal air changes through ventilation will help to alleviate this condition but running costs together with maintenance and heat losses will also have to be considered.

The application of external insulation of sufficient standard will assist in allowing the increase of the wall temperature above the "dew-point" thus helping to prevent wall surface condensation forming and reducing the likelihood of the formation of fungus growth.

When the external insulation system is installed correctly the existing wall remains dry and the temperature is raised more easily and efficiently by the internal heating methods.

There are no running costs associated with the insulation method and periodic maintenance should be low if installed to a reasonable certified standard.

The advantages of External Insulation can be itemised as follows:-

1.   Raises/lowers the temperature of the wall which will:-

     a.   Reduce or eliminate internal condensation.
     b.   Preserve the structure from decay.
     c.   Keep the structure dry.

2.   Provides a new or revitalised appearance to the external finish

3.   Reduce Heating/Air Conditioning Costs due to:-

     a.   Increased insulation to modern standards.
     b.   Reduce decoration costs due to reduced condensation levels.
     c.   Even-out thermal cycling thus improving comfort levels.
     d.   Reduced draughts through improved building details.

4. Improved Internal Comfort Levels for the Occupants due to:-

    a. Improved heating/cooling reaction times.
    b. Lower internal heat gains in summer conditions.
    c. Reduced or eliminated water penetration through walls.
    d. Reduced sound transmission.

1.4 Terminology and Definitions

The term "External Wall Insulation" literally means the insulation of the "External Wall". However, within the industry, this term specifically refers to the application of an insulation system to the external face of the wall.

Recently the trade association known as "The External Wall Insulation Association" or E.W.I.A., changed it's name to "The Insulated Render & Cladding Association" or INCA, in order to be more specific in it's definition as to the roll it plays in representing the organisations and company members within this industry.

Common definitions used in this application are as follows:-

Accessories Plastic or metal beads, trims or profiles which are used to complete a system at edges, abutments and which help control details and are weather-tight to complete a durable installation. This is to include Copings and cills.

Adhesive A pre-mixed compound used to affix the insulation board or any other item to the substrate in accordance with manufacturers instructions.

Acrylic Renders & Finishes A render or textured finish with acrylic additives to improve waterproofness and flexural strength

Basecoat A waterproof compound applied directly to the insulant to be reinforced as necessary for the particular application in order to support the final finish.

23

Board Claddings   Thin board materials of timber, cement or metal compositions which will clad studs and incorporates a self-finish.

Bond Breaker   A material that prevents adhesion at a designated interface in a sealant joint.

British Board of Agrément (BBA)   The British Board of Agrément is one of the principal testing authorities for cladding and insulating systems in the UK and are based at Watford. See Part Six for contact details.

Brick Slips (Clay)   Thin facings bricks 10-20mm thickness to be used as a cladding to represent a traditional brick wall when pointed in the conventional manner. Available in a wide variety of colours, textures and sizes throughout the UK.

Brick Slips (Resin)   Thin facings bricks 5-8mm thickness used as above fixed directly onto a polymer modified reinforced basecoat which is used as pointing.

Building Research Establishment (BRE or Wimlas) . BRE Certification Limited is one of the principal testing authorities for cladding and insulating systems in the UK and are based at Watford. See Part Six for contact details.

Coatings   Bucket supplied in pre-blended colours and texture sizes, these coatings cover renders to provide protective decorative finishes.

Dash Receiver   A render applied over a basecoat suitable to be applied with aggregate chippings to create a natural stone finish. Can be applied in a variety of colours and thicknesses.

Dry Dashes   Natural or manufactured aggregates to be thrown onto wet cement renders, known as dash receivers, to create a finish. Available in a wide variety of colours, sizes and textures.

External Wall Insulation System (EWIS)   A non-load bearing exterior wall cladding system comprising an insulation board, adhesive or mechanically fixed to the substrate, an integrally reinforced waterproof basecoat together with suitable beads and trims, completed with durable decorative applied protective finishes.

Finish Coat   The final application of the finish to provide a decorative texture or protective finish.

Fire Resistance   The rated ability to resist fire propagation from an external source

Fire Spread   The rated ability to resist the spread of fire from an external source.

Fire Break   A protective break in a system to help prevent the transfer of fire, from an external source, particularly in multi-storey buildings.

Fixings or Fasteners   Corrosion resistant assemblies of various configurations used to secure systems to the substrate by pre-drilled holes. Usually supplied in plastic or stainless steel.

INCA   Insulated Render & Cladding Association the Trade Association for the industry, see section six for full contact details.

Insulation   A slab material that provides thermal resistance to reduce heat flow through the wall. This slab also provides, in render systems, the backing to carry the waterproof layer and decorative finish.

Insulated Render   A render material, usually cementitious in nature, with an added insulating material to increase its effectiveness to resist the thermal transition of heat or cold. Applied by hand, requires a finish to be applied on completion.

Overcill   A new cill over-cladding the existing cill to cap the new insulation system.

Polymer Renders Cementitious renders with additives of polymers to increase flexural strength and waterproofness. Available in a wide variety of colours when used as a top-coat, natural in colour when used as a basecoat.

Primers Specialist paint-type products used to enhance system performance for adhesion or water resistance. Certain primers are used to condition the substrate prior to the application of adhesive, other primers are primarily used to enhance the security and water resistance of textured coatings.

Renders - Cementitious Traditional sand/cement renders usually mixed in accordance with BS5262. These renders are usually applied onto a metal carrier, are waterproof and are finished in a wide variety of colours.

Renders - Acrylic Pre-mixed and supplied in a bucket, these renders are supplied in a wide variety of colours and textures. Specialist application onto pre-prepared backgrounds.

Renders - Silicone Pre-mixed and supplied in a bucket, these renders are supplied in a wide variety of colours and textures. Specialist application onto pre-prepared backgrounds.

Reinforcements A mineral fibre or metal material used to strengthen the render finish for impact resistance and durability.

Sealants A material that has the adhesive and cohesive properties to form a barrier against the passage of liquids, solids or gases.

SBR Additive Styro Butyl Resin, used to modify mortars to increase adhesion and waterproofness.

Silicone Renders & Finishes A render or textured finish with silicone additives to improve waterproofness and flexural strength

Simulated Brick Finish An applied render of a variety of colours, suitably marked-out into brick-courses to simulate conventional traditional brickwork.

Simulated Stone Finish An applied render of a variety of colours, suitably marked-out into random-courses to simulate conventional natural stonework.

Substrate Building substrates are found in a variety of materials such as brick, blocks, stone and many sheathings attached to timber or steel studdings.

Terracotta Tiles A system of dry cladding with insulation combined to provide a composite finishing system of architectural interest.

Tile Hanging A system of hanging tiles in a vertical situation from timber supports. The cladding can be insulated with the addition of insulation secured between supporting battens.

Tiles can be supplied manufactured in clay or concrete, to a variety of sizes, colours and profiles.

Trims & Beads Supplied in metal or uPVC to complete the render at edges and abutments.

Undercill An new cill installed with the system to extend the existing cill to cap the insulation and render finish.

Verge That part of a roof that is the edge of a gable.

1.5   System Types

External wall insulation systems generally comprise an insulation layer protected with a weatherproof finish, usually a render, but can comprise brick slips, tiles and decorative boards. Insulation boards are required to achieve the requisite standards of thermal performance, the finish to provide the protection and decorative finish. Insulating render can also be an advantage in certain locations.

Board Systems - Wet Finishes

Insulation boards can be namely:-

Expanded Polystyrene Boards
Extruded Polystyrene Boards
Polyisocynurate or Polyurethane Foam Slabs
Phenolic Foam Slabs
Mineral Fibre - Rock or Glass Rigid Slabs
Foamed Glass Slabs

Typical Traditional Render Finishes

Renders can be   Heavyweight
                      Medium-weight
                      Lightweight
Reinforced with
                      Mineral fibre mesh
                      Polypropylene mesh
        (medium/lightweight renders)
                      Galvanised mild steel mesh
                      Stainless steel mesh
        (heavyweight render systems)

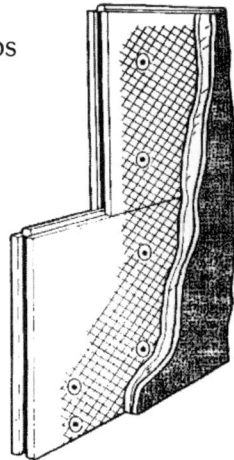

Fixing types & sizes will depend on the substrate and design exposure requirements. Finishes will be to the requirements of the client or planning officer but generally will include the following:-

Render finishes comprise
> Dry-dash Aggregate Chips on coloured polymer renders
> Scratch Render Finishes
> Roughcast render finishes
> Simulated brick finishes
> Simulated stone finishes
> Smooth Render Finishes
> Textured Coatings
> Painted smooth render

Board Systems - Dry finishes
Cedar Timber boards
Vertical Clay/Concrete Tile Hanging
Terracotta tiles
Aggregate Finished cementitious boards
PVC Ship-lap Boarding
Timber (sw) Ship-lap Boarding
Aluminium profiled claddings (Painted)

These dry finishes are usually fixed to the substrate by means of timber battens independently fixed to the substrate, insulation between the battens and the finishes fixed to the battens.

Insulated Render Systems   Insulated renders are cement based renders containing expanded polystyrene beads and other additives to provide a render which will perform in the conventional manner but also provides a degree of insulation. The thickness when applied, will govern the thermal performance. A wide variety of textures and colours and finishes can also be provided.   There are certain advantages with this method, particularly when insulation boards are difficult to fix or when penetration of the render is frequent and complex.

Whilst the render has to be applied in thicknesses greater than normal renders to be effective, the insulated render does solve those problem areas where boards are difficult to apply, or where only very thin boards are practical and the final thermal values are difficult to achieve, particularly at the reveals of windows and doors.

29

Cavity Drainage or Ventilated Systems Conventional board insulation systems with a designed cavity between the render finish and the insulant. This cavity can be formed in a variety of methods namely :-

1.   Profiled Expanded Polystyrene with channels at the interface with the render finish.

2.   Composite cement-particle board, bonded in two layers to create a cavity.

3.   Board insulation systems with dry-applied finishes such as tile-hanging, aluminium boards and terra cotta systems.

1.6   *Traditional System types currently in use (Selection of)*

Rockwool Insulation
Medium weight render with scratch/dash finish

Phenolic Foam Insulation
Medium weight render with scratch/dash finish

Expanded Polystyrene/Rockwool Insulation
Light weight render with textured coating finish

Expanded Polystyrene/Rockwool Insulation
Heavy weight render with scratch/dash finish

Note :- All of the above systems can have the complete range of finishes applied to them and to include :-

Acrylic/silicone texture finishes,
Simulated brick renders,
Brick slips,
Dry claddings

Table of Performances Table of Insulation Requirements for various wall values to achieve current Building Regulation requirements of 0.3W/m²DegC.

Assuming a standard external render finish thickness of 10mm

Uninsulated Wall Values W/m²DegC	Insulation Type	Thickness	Render
2.00	EPS	80-90mm	10mm
	Rockwool	80-90mm	10mm
	Phenolic Foam	50-60mm	10mm
1.75	EPS	75-85mm	10mm
	Rockwool	75-85mm	10mm
	Phenolic Foam	45-55mm	10mm
1.5	EPS	70-80mm	10mm
	Rockwool	70-80mm	10mm
	Phenolic Foam	45-55mm	10mm
1.25	EPS	65-75mm	10mm
	Rockwool	65-75mm	10mm
	Phenolic Foam	40-50mm	10mm
1	EPS	60-70mm	10mm
	Rockwool	60-70mm	10mm
	Phenolic Foam	35-45mm	10mm

A table of values demonstrating the necessity of selecting the correct insulant for the project as thicknesses will be important in detailing the project correctly to avoid water ingress and produce a visibly aesthetically pleasing finish.

The total thickness of the ultimate system may also be influenced by the method and type of fixing permitted by the exposure and substrate type and quality.

This is a guide only, calculations should be evaluated on a job specific basis to confirm values and requirements.

1.7   Traditional Finishes

The following is a selection of traditional finishes currently utilised within the external wall insulation industry.

Dry-Dash Render   Traditional render application used throughout the whole country, dry dashing aggregate is thrown onto the wet render to create a natural aggregate finish. Available in a wide variety of colours, sizes and textures, the practice is relatively cheap, with skills widely available. Manufactured aggregates such as ceramics and glass are also available and are used for the more specialised projects as the supply costs are considerably greater than for natural aggregates.

Scratch Plaster - Render   Coloured cementitious renders are scratched with a scratching tool whilst the surface is still workable but after the initial set has taken place. The surface of the render is removed by the action of the scratching tool and approximately 2-3mm of render are removed exposing the open matrix of the aggregate mix. The true colour of the render is exposed with a light even texture which is visually very pleasing.

Rough-cast Render   The top-coat render and aggregate mix is thrown onto a backing coat in a slurry form, the aggregate totally encapsulated within the cementitious slurry. The aggregate is any hard stone of an equal graded size to suit the particular application and creates the final "lumpy texture" finish. Traditionally widely used in Scotland.

Silicone Enriched Renders   A new innovation is the inclusion of silicone water-proofers in pre-blended and pre-packed proprietary renders. This development increases the specification and capabilities of polymer renders, particularly for exposed or coastal areas. Applied in the conventional manner and now readily available in all the usual colours.

Tyrolean Finishes (cementitious)   A form of sprayed cementitious mix, pre-coloured and applied by a hand-held machine, this finish is widely used throughout the UK as an economical, easily applied colourful finish for all forms of building type. Medium term durability under average conditions.

Smooth/Painted Finishes Masonry paint applied in accordance with the manufacturers instructions, this finish is applied to a good rendered surface to give a smooth coloured effect, free of imperfections with the more natural aggregate finishes can sometimes deliver. A very wide selection of colours, usually light in choice are available.

Textured Coatings Applied by roller or trowel to an approximate thickness of 1.5-3mm, these coatings are usually acrylic or silicone based for waterproofness and long term durability. The final effect is with an even but flat textured finish.

Acrylic variant :- easily applied and initially "high performance" appearance finish, disadvantages can include a tendency to loose opacity and brightness of finish over a relatively short life span. Acrylics provide a medium "sheen" finish.

Silicone variant :- more resistant than acrylics to marine environments as they have superior water resistance but can be more costly.

Brick slips Brick slips are a thin facing or part of a brick that when applied over an insulant provides a "brick" wall finish that is easily recognisable as a traditional brick wall finish.

> Usually supplied 5-8mm thicknesses - in resin,
> 10-20mm thickness thicknesses - in clay
> 20mm + Cut slips

these slips can either be extruded manufacture or cut slips from real bricks. Coloured pointing in specialist waterproof pointing mortars complete the desired effect.

Commercially available brick slip systems are developed as a complete system to include tested means of support for the brick slip finish, certified by BBA/BRE.

Simulated Brick Renders Brick finishes can be replicated in coloured polymer renders to a high standard of visual acceptability. Two coloured layers of polymer modified external cementitious render are applied in thin 3-4mm layers onto a specified coloured backing. The "brick" pattern is cut into the top layer exposing the under-layer of differing colour representing the cement joints. Any desired brick-bond convention or size of brick can be replicated.

Simulated Stone Renders Natural stone finishes can be replicated in coloured polymer renders to a high standard of visual acceptability. The principal is that two coloured layers of polymer modified external cementitious render are applied in thin 3-4mm layers onto a specified backing. The "stone" pattern is cut into the top layer exposing the under-layer of differing colour. This under-layer will represent the cement joints and can follow any desired convention or size of stone. Normal design restraints as to conventional external wall insulation applies.

Terracotta Tiles Terra cotta tiles are usually clay burnt material imported from Europe supplied in the usual terra cotta tile colour. These proprietary systems are secured by a system of extruded aluminium rail systems fixed securely to the substrate with insulation material inserted within the void. Special pressed metal profiles and cills complete the systems to make them waterproof. Available commercially as a complete system.

Board Finishes Timber boards, aluminium or PVC "sidings" are installed over insulation to provide the architect with additional alternatives to the traditional render and applied finishes. The boards are usually of Ship-Lap profile and supplied in Cedar or Softwood (treated), Cedar is for a natural finish the softwood being painted.

Commercially available aluminium or PVC systems are available for consideration as alternatives for projects where these finishes are considered appropriate.

Tile Hanging Traditional tile hanging finishes can be applied onto conventional timber support structures and insulated with the required type and thickness of insulation to the demands of the building. Suitable consideration to be given to ventilation passages to avoid interstitial condensation. Many sizes, designs and colours are available in these tiles and are suitable for use with insulation.

Historic Lime Renders Certain areas of historic importance, whereby local buildings have been rendered with natural lime renders may be required to replicate such historic finishes by the Planning Authorities. Lime renders can be provided as a finish over a suitable base render to facilitate such a finish, specialist manufacturers and suppliers must be consulted for the appropriate specification installation and trained specialist applicators.

Rain-screen Cladding   Proprietary rain-screen claddings include high levels of insulation to comply with current Building Regulations.   The systems are usually fabricated using extruded aluminium sections with insulated panels to cover complete elevations providing a new overcoat which will substantially change elevational appearances as well as improving the overall building's thermal performance.

A specialist supplier providing a good provision of detailing, quality of manufacture, accurate installation with good workmanship is essential.

1.8   Selecting a System and Producing a Contract

The supervising Surveyor, Architect or Project Manager must consider many factors in the selection of an External Wall Insulation System to meet certified standards and comply with all the necessary Planning and Building Control legislation.

The system must have been tested and certified by the British Board of Agrément and/or BRE Certification Ltd. These organisations set the standards of acceptability within the UK for normal use with standards of durability and fire performance.

External Wall Insulation System suppliers are recommended (optional, not compulsory) to be registered with the Insulated Render & Cladding Association, the Industries' Trade Association. The thermal standards, system construction and finishes being selected by the specifier.

"It is easy to apply a system on a flat wall!" an easy statement to make, however, all projects will have a problem somewhere. Careful consideration will be necessary to avoid future long-term difficulties, the system will require to be installed with junctions, abutments, movement positions and many other considerations requiring thoughtful detailing. The correct use of trims, beads and accessories are an important factor to ensure durability and certification for the benefit of the Building Owner as well as for the protection of the installing Contractor.

The following is a "check-list" of elements of systems which primarily make up the contract or project, they are:-

1.   Thermal standards to be achieved knowing the existing building performance which will determine the quantity of insulation to be provided.

2.   The fire performance to be achieved, with due consideration to the height of the building and its use.

3.   The security of the system to withstand exposure to the

elements, either wind or rain, due to height above sea-level and closeness to shores etc..

4.   The detailing necessary to achieve a satisfactory installation both visually and economically.

5.   The degree of wear and tear likely from occasional impact from the normal environment of conventional living to the possibility of deliberate damage from vandalism will decide the robustness of the system to be installed.

6.   The finish of the system to be installed, taking notes of local materials, desires of Local Planners and satisfying the wishes of the local population and owners/occupiers.

7.   The system to be economical and able to be contracted within desired budgets.

Contractors are notoriously difficult to control both from the economies of the project to the standards of workmanship delivered on completion. They are recommended to be registered with the Insulated Render & Cladding Association (INCA) who offer an inspection of workmanship system together with a financial stability standard and "Code of Conduct". Grievance procedures are also handled within this organisation.

A typical contract calls for the Contractors to be assessed before invitation to tender, the following items need to be considered :-

1.   Check membership of INCA, the requisite Trade Association.
2.   Check financial standing of all the Contractors to be invited to tender.
3.   Check workload of the Contractor and the ability to complete the contract.
4.   Take up references if deemed necessary.

5.   Examine previous completed contracts.

The following "check-list" is to outline the necessary procedures to compile a project and satisfying all of the necessary requirements with any interested Statutory Bodies, before contract.

1.   Visit the site and survey elevations of properties to determine architectural features, difficult detailing and any other associated problems.

2.   Survey site to determine contractual responsibilities.

3.   Check certification of proposed system (BBA or BRE)

4.   Test walls for any defects to existing walls claddings or finishes.

5.   Test walls for "pull-out" loads to assist in the evaluation of fixing types and quantities.

6.   Check that any substrate variations are compatible with the system of trims to be used or, whether or not special trims are required.

7.   Evaluate access, heights of buildings, security of tenants etc..

8.   Contact local Planners to evaluate any planning problems.

9.   Contact the local Building Control Officers to evaluate any requirements as to any specific local requirements.

10.   Discuss finishes, colours and textures with any interested party such as owners/occupiers.

11.   Evaluate any service problems such as overhead electricity, TV, telephone cable supplies which may have to be repositioned.

12.   Discuss any waste disposal problem with the Local Authority.

13.   Discuss any site access with the Local Planners, Land Owners or Traffic Police, to avoid local congestion or nuisance.

1.9 Costings and Comparisons

The cost of external wall insulation and the "pay-back" as a result of fuel savings has never been significant with the application of such systems. However, the following should be considered in addition to the initial cost.

Improved comfort levels The application of external wall insulation improves the thermal cycling within the building, eliminating severe temperature fluctuations and thus improving comfort levels. This phenomena is very difficult to value in monetary terms and is only really appreciated by experience.

Reduced or eliminated condensation The external wall insulation increases the external wall temperature above the dew-point thus eliminating or reducing condensation. Values of this improvement can only be registered by reduced re-decorating and maintenance costs.

Protection of the structure The application of external wall insulation protects the structure it is applied to and so extends its serviceable life. Walls have reduced maintenance, less weather exposure and overall reduced wear-and-tear, thus adding to the value of the cladding process.

Improved appearance of the building The new final finish of the external wall insulation system will obviously look clean and fresh giving the building a re-vitalised appearance. The new visual impact will almost certainly add value to the property.

Improved sale-ability of the building The overall cost of the external wall insulation cladding system, combined with all of the above improvements will significantly improve the sale-ability of the property by making it more visually desirable over un-treated properties together with its increased energy performance.

Future escalation of fuel costs Energy costs during the latter part of the 20th century and the beginning of the 21st, are increasing at a steady rate without any indication of a slow-down. The facts are that the increasing dependence on foreign energy supplies, the poor acceptability of natural forms of energy generation and the likely-hood of a return to massive nuclear energy production will continue to increase the cost of fuel. Political reliability on energy supplies are also becoming more uncertain. The philosophy of increasing the tax on fuel to reduce consumption is also a possibility.

With all of these possibilities for the future, the only certainty is that fuel will increase in cost with the result of increasing the demands for more energy efficient buildings and heating/air conditioning systems. External wall insulation will therefore become increasingly more popular by the resultant better use of available fuels. The cost comparisons will continue to improve as fuel becomes more expensive. In simple terms, as fuel costs increase the pay-back period for the installed insulation system decreases.

Energy Efficiency Rating   The proposed new Energy Efficiency Rating recently announced by Government, will consider the building's overall thermal performance. The installation of external wall insulation will improve this rating which will subsequently make the property more attractive to a buyer and so increase its value.

Air Conditioning   During periods of high external temperatures, air conditioning becomes a reality in preserving comfort levels. The use of external wall insulation will reverse the building's thermo-dynamics and assist with the preservation of cool air within the interior.   As the cost of air conditioning relative to heating is substantially higher, the savings made by the introduction of external wall insulation will be greater (dependent on local circumstances and standards)

1.10 Planning & Building Regulations

Contracts incorporating external wall insulation within the UK are usually in two forms, namely "new-build" and "refurbishment". The "new-build" contracts will require planning permission together with Building Regulation Approval in the conventional manner. Whilst "refurbishment" contracts may only require these approvals if associated with other works directly attributable to the external wall insulation installation.

If in doubt, contact the relevant Planning Authority for advice and assistance.

Planning Permissions

Installing external wall insulation (within the UK) may significantly alter the elevational appearance of the building and as such may require planning permission. This is particularly relevant if the building is situated in the following designated areas :-

1.   Conservation Area

2.   Area of Outstanding Natural Beauty

3.   National Parks Authorities

Additional considerations must be given to elevational treatments if the project involves a "Listed Building" consent.

Building Regulations

The current UK Building Regulations apply specifically to "new-build" projects but may apply to "refurbishment" projects if there is associated other work. If in doubt contact the local Building Inspector for help and assistance.

Current requirements are for a "U" value of $0.3W/m^2DegC$, for "new-build" which is also the "target" value for normal refurbishment projects. Building Regulations Part L1A provides for :-

Limiting "U" Value Standards (W/m²K)

Element	Average "U" Value	Limiting "U" Value
Wall	0.35	0.70
Floor	0.25	0.70
Roof	0.25	0.35
Windows, roof windows rooflights and doors	2.2	3.3

Building Regulations Part L1B provides for :-

Standards for Thermal Elements (W/m²K)

Element	New Thermal Elements in an extension	Replacement Thermal Elements in an existing dwelling
Wall	0.30	0.35
Pitched Roof Insulation at ceiling level	0.16	0.16
Pitched Roof Insulation at rafter level	0.20	0.20
Flat Roof or roof with integral insulation	0.20	0.25
Floors	0.22	0.25

Notes

The building fabric should be constructed so that there are no reasonably avoidable thermal bridges in the insulation layers caused by gaps within the various elements, at the joints between elements and at the edges of elements such as those around window and door openings.

PART TWO - System Performance

2 System Performance & Criteria

An External Wall Insulation System has to protect and insulate by combining the many components into an integrated insulation system. It is required to support its own weight, waterproof and insulate the wall and be sufficiently durable to withstand the rigours of many years of extreme climatic variations and changes.

The complete installation should be designed in accordance with all relevant Certifications, British Standards, Codes of Practice and INCA recommendations. The final appearance shall offer an equal or improved appearance over the original and will not be detrimental to its surroundings or environment.

The thermal performance shall be targeted to achieve current codes and Building Regulations.

Various organisations have brought together EU members to develop common standards for insulated cladding systems. These European Union of Agrément (UEAtc) established ETAG 004 (previously to the requirements of M.O.A.T. 22) which measures the system performance using the latest specified tests. This criteria is used by the British Board of Agrément (BBA) and/or BRE Certification These standards are a valuable resource for Architects and Designers in setting their own requirements of project design and management.

The following are a resumé of some of the various testing procedures carried out by the relevant testing authorities, BBA or BRE, to enable them to certify compliance with the above European regulations and standards.

1. Hygrothermal cycles test -
   - Heat - rain cycle
   - Heat - cold cycle
2. Freezing tests
   - Freeze/thaw cycles
3. Capillarity tests
4. Depression resistance
5. Bond strength tests
6. Pull-out tests of fixings
   - Pull-out from wall
   - Pull-through the system
7. Cohesion test
8. Compression test
9. Shear modulus of elasticity test
10. Dimensional stability test
11. Behaviour on exposure to water test
12. Bond tests for adhesives
13. Testing after aging
14. Tests on components
15. Hygiene, health and the environment
16. Safety in use
17. Responsibilities
    - Factory production control
    - Initial type testing
    - Continuous surveillance of production
18. Documentation
    - Basic manufacturing process
    - Product and material specifications
    - Test plan

The following items will highlight the basic requirements for an efficacious installation.

2.1 Safety

Concerns for the public safety and welfare require that the External Wall Insulation System is designed to provide a secure attachment and reasonable fire-resistance. Unsatisfactory safety performance can be life threatening due to the danger of falling materials.

Any external wall insulation system supplied and installed to any building should perform its function by remaining intact, without the creation of any risks to the public or to any person associated with the use of the building, by whatever cause.

All elements of the installation shall be secure and not subject to any risk of failure from any imposed external fire spread.

Attachment   The system should be a fully integrated and tested assembly, the interface of the varying layers must be well bonded or fixed with the complete system secure to the substrate. Include the following considerations:-

1.   Adhesives, mechanical fixings or a combination of both, should provide adequate safety margins under expected reasonable normal working conditions.

2.   The system should be capable of withstanding reasonable thermal or structural movement within the supporting structure or substrate without sustaining failure in the form of cracking, deformation or detachment.

3.   Attachment should resist the combined load of its own weight and laterally induced loads equivalent to those produced by job-specific wind pressures and exposures.

4.   Resist the results of impact damage, by remaining safe to all users or the public, whether accidental, by natural causes or by vandalism.

The Victoria Centre Nottingham, illustrated above is a perfect example of the necessity for security of fixing, being twenty-nine storeys high with the consequential severe exposure to wind and weather.

References	*1*	*Certification*	*- Section Two, item 2*
	2	*Fixings*	*- Section Three, item 3.3*
	3	*Beads-Trims*	*- Section Three, item 3.10*
	4	*Fire*	*- Section Four, item 4.7*

Fire Resistance   The system and its individual components should not contribute to any significant spread of fire from within or without the building or structure, including ignition from radiant or conducted heat, so that the external wall insulation system   can continue to perform its intended functions in the event of such exposure. Adequate fire-breaks or compartments should be designed into the project where   deemed necessary or as required by the Building Regulations.

Many adhesive fixed systems rely on thermoplastic insulants such as expanded polystyrene to support the final finish, therefore if the insulant has cause to soften by heat, the exposed finish has no direct support as expanded polystyrene has no strength above 100degC. Once the finish is partially detached it can oscillate in the convection currents above the fire. This leads to a pumping action which can suck hot flame and gases behind the finish and extend the area of damage.

Above three storeys, a "fire-safe" insulant with "fire-proof" fixings and fire-barriers at suitable levels is recommended.

The fire resistance of the render or finish should also be considered and have at least a "Class 1 Surface Spread of Flame" to BS 476 with a Class O composition.

Fire Resistance to Walls   The external wall insulation system, as installed, shall not increase the fire risk to the building, its occupants or its surroundings.

All compartment walls within the building may be extended through any system to the outside face, with the provision of fire-breaks, so preserving the original building design in respect to compartments.

The head or shanks of all plastic fixings may be heated sufficiently in a fire to release the cladding which can lead to excessive damage. To resist this, a combination of fire resistant fixings should be considered along with the fire resistant performance of the insulation and render.

A safety first policy is to adopt on any high rise property, only a "Fire Safe" insulation material, thus reducing the necessity for fire breaks within the insulation system.

Fire Resistance in Multi-Storey Buildings   There is no particular requirement for any special provisions for low-rise buildings, up to three storeys, but above this the action of fire both internally and externally must be considered.

Fire can penetrate behind insulated claddings, which are insufficiently stabilised, causing a vertical spread of fire or chimney effect, up to many storeys. To help prevent this, the inclusion of fire resistant fixings, "fire-safe" insulation materials and incombustible fire-breaks is essential.

Compartment floors in multi-storey building should be replicated through the insulation system with the use of incombustible fire-breaks or additional fire-breaks at every two storeys to help resist or contain the transmission of fire vertically.

The present convention to provide fixing security in case of fire on multi-storey buildings is with the provision of stainless steel anchors at the minimum rate of one per m² of wall area.

Fire resistance & Drained System Some projects, particularly those in exposed locations, may require a form of drainage to protect the insulant in particular and the whole wall in general.

This can be provided in a variety of ways with the formation of cavities within the system. The design and construction uses specifically designed profiled polystyrene or dry claddings.

Particular attention is drawn to the design of a drainage system which may require the provision of "fire breaks" to prevent the passage of fire vertically.

References	1	Fire	- Section Four, item 4.7
	2	Inspections	- Section Five, item 5.16

2.2 Thermal Performance

The thermal transmittance or "U" value of a wall of a building is a measure of its ability to conduct heat out of the building; the greater the "U" value the greater the heat loss through the structure.

The calculation of "U" values is generally in accordance with BS 5250 with the additional calculation of condensation risk to ISO 13788.

The present convention associated with external wall insulation does not require any reference to heat loss through windows or other openings, except when considering the total "envelope" of the building.

The advent of S.A.P. (Standard Assessment Procedure) ratings to determine the total building's thermal performance will require all windows, doors, roofs, floors and walls to be considered together with other energy producing elements to produce a complete package. The provision of external wall insulation will significantly help to increase the SAP rating and so meet Government targets on insulation performances.

Cold bridges through the existing structures are to be avoided wherever possible due to possible condensation risks, but again there is no convention to bring any cold bridge into calculations unless S.A.P. ratings are included.

The specified thermal performance of any system shall provide the building owner with a thermally improved weather-tight wall, within reasonable cost constraints, provide current legally required thermal resistance values without associated problems such as condensation and unstable finishes.

The current method of "pinpoint" "U" values through walls now has improved to a calculation based on a S.A.P. rating for the building as a whole. This requires calculations taking values through each varying substrate and aggregating them, hence "whole-envelope calculation".

| *References* | 1 | *BRE Digest 190* | *Heat losses from dwellings* |
| | 2 | *BS 5250 1975* | *Control of Condensation* |

2.3   *Wind Loadings*

The system shall be designed to withstand the normally expected actions of wind, anticipated under the general parameters of CP3, Wind Loadings. *(Code of Practice no3)*

Pull-out test on"No-fines"structure.

From British Standards CP3 Chapter V part 2: 1972 the likely effect of various wind velocities on a building in the UK can be ascertained.   Maximum design velocities vary from 38 metres per second in the London area, up to values in excess of 54 m/s. elsewhere.

A further factor has to be included depending upon exposure, in particular the height of the building and it's elevation above sea level. Sites close to the coast must be considered as having a "high" exposure for wind loadings.

Generally the greatest stress is created by the degree of suction which occurs close to the upper corners and edges of a building. For a building of severe exposure and with a wind speed of 54 m/s the amount of negative pressure is in the order of 300kgf/m^2 (3,528 N/m^2).

It must be concluded that although mechanical fixing at close centres is not required to carry the weight of the system it is safer to

50

ensure that the full quota of pins is specified so as to maintain resistance against wind forces.

The performance of the fixing pins is subject to the results of a "pull-out" test to determine the quality of support of the substrate. Differing substrates will offer differing values of "pull-out" which will determine the number and type of fastener to be used. The "pull-out" test will also determine the minimum depth of embedment within the substrate, usually the minimum recommended depth is 50mm to allow for any deformation of the entry to the drilled hole and sleight variation in drill sizes.

In addition to the pull-out performance of the fixing into the substrate, the fixing must also resist the "pull-through" by the action of wind suction. This is the ability of the fixing to remain intact avoiding any possibility of the system being "forced-off" the fixing by wind suction. The spacings of the fixings may contribute to this overall performance by reducing the effective span of the system between fixing points.

References   *1*   *BRE Digest 119*   *Assessment of Wind Loads*

2   *BS Code of Practice No 3 - Wind Loads*

3   *Construction Fixing Association :-*
Guidance note for Brickwork & Block
European Technical Approvals for Construction Fixings

2.4   *Condensation Risks*

The avoidance of the production of condensation, both internal and interstitial, must be considered in any externally insulated wall project. Condensation is the formation of water out of natural air after reaching its "dew-point" causing damage to wall surfaces and materials where they are susceptible. Condensation internally may also be responsible for the poor health of occupants as a result of mould-growth.

Condensation occurs when warm moist air meets a cold surface, whether this is visible or invisible, internal or external or within the structure of any wall. Condensation is more likely to occur in winter conditions when the building structure is cold, the internal air temperature is warm and in locations where the Relative Humidity of the internal air is very high for short periods. i.e in kitchens and bathrooms.

Ventilation systems together with heat exchanger/recovery units, can determine the extent or elimination of condensation. The installation of mechanical fans, chimneys, ducts and window trickle vents is now a requirement through Building Regulations for new buildings and extended buildings to kitchens and bathrooms.

Good detailing is essential to eliminate condensation through "cold-bridges" i.e. where the wall structure is continuous without insulation thus providing a direct link for thermal movement to the external air.

Interstitial condensation occurs within the substrate i.e. between the layers of differing materials of the structure of the wall and is not visible, albeit the long-term effects will ultimately become visible through the transmission of damp. Condensation, interstitial and internal, formed when the "dew-point" is reached can be eliminated with the installation of the correct type and thickness of insulation layers.

References   1   *BRE Digest 110*   *Condensation*
              2   *BS 5250 1975*   *Control of Condensation*

2.5   *Impact & Puncture Resistance*

The system should be capable of resisting deformations or damage arising from hard and soft-body impacts. It should also be capable of resisting damage following impacts caused by normal wear and tear and/or vandalism however caused.

Varying degrees of impact can be anticipated dependant on exposure, location and usage. The systems available can be varied and adapted to suit these circumstances which can be graded as follows:-

1.   Areas of walkways and public footpaths close to buildings.

2.   Play areas and playing fields close to buildings.

3.   Protected areas such as gardens or landscaped areas.

The above does not include any allowance for the acts of vandalism which can, of course, be incurred in any location or social environment and varies in severity.

Test laboratories carry-out experiments to determine impact or puncture resistence. The hard body impact load with steel ball and dynamic indentation with Perfotest represents the action from heavy, non-deformable or pointed objects which by accident hit the system. The test comprises a steel ball dropped vertically in controlled conditions through a 1m tube onto the complete system.

1.   Superficial damage, provided there is no cracking, is considered as showing "no deterioration".

2   "Penetrated damage" if circular cracking is observed which reaches up to the insulation.

3   "Perforated damage" if destruction of the rendering is shown up to a level beneath the reinforcement in at least 3 of the 5 tests.

The results of impact damage to the insulation system will ultimately cause deterioration in performance, appearance and may impair its durability.

Damage severe enough to penetrate to the insulation layer will cause the performance of the system to be severely thermally reduced if exposed to weathering for long periods of time.

Remedial action is desirable to reduce the long-term effects of any thermal reduction, good visual repairs are desirable to ensure maintenance of appearance and reduce any decline in overall performances.

References	*1*	*Render design*	*- Section Four, item 4.5*
	2	*Protection*	*- Section Five item 5.15*
	3	*Impact Damage*	*- Section Five item 5.18*

2.6   Temperature Tolerance

The system should not suffer irreversible damage or deformation due to the extremes of cold and heat.

Typical extremes of temperature for the UK are:-
Low -15DegC
High +35DegC

The higher temperature will result in a significantly higher surface temperature due to solar gain, these surface temperatures can be as high as +50DegC. The insulated render will have to perform satisfactorily across this temperature range without deformation or cracking to provide long-term durability. The thinner the render and finish or the darker the colour, means that the temperature range will be greatest as there is little mass to absorb the heat gained.

Under certain circumstances, particularly in winter conditions, the temperature gain can be very rapid as frosty nights can turn quickly to sunny daytime, this effect is known as "thermal shock". The low mass of the insulated render has no ability to store heat and so any imposed heat is quickly lost, the "thermal shock" is equally prevalent for the cooling effect as well as the heat gain effect.

The system has to sustain movement by "thermal shock" in all directions and in all locations, where fixed point locations are found, such as corners of buildings, windows and door openings, the render will have to move within these points. Movement control joints may be installed at strategic locations to blend visibly and satisfactorily with elevational treatments. Any movement joint installed should have adequate design to cater for any "thermal shock" which may be generated together with being adequately weather-proof to sustain extreme weather conditions such as driving rain.

The selection of final colours has also to be considered as they can contribute to movement due to "thermal shock", it is well known that dark colours absorb more heat than light colours resulting in differing movement patterns on similar sized projects. It may be that on some work of a similar size where differing colours are required by the Architect, varying movements must be allowed for.

Thermal stresses can be relieved by increasing the number or size of movement control joints, sealants may also have to be considered and the correct size and quality of a sealant joint may be necessary to assist or compliment the former.

Movement joint positioning is dependent on the format and layout of the building to be insulated or rendered, the only constant being that existing structural joints should be continued through the external wall insulation system. If careful consideration is given to movement joint positioning, these can effectively visually "disappear" into the well designed elevation. They can also be used to act as a "marker" for a change in materials or colours. Movement control joints can be sighted behind rainwater pipes or plumbing ducts to become inconspicuous to the general scene.

It is noted that some lightweight external wall insulation systems incorporate renders or coatings sufficiently flexible to avoid the use of movement control joints. Specialist advice from the respective specialist system provider should be sought if this is a particular requirement by an Architect on a project.

References   *1*   *Render Design*   *-Section Four, item 4.5*

2.7   Moisture Protection

The external wall insulation system, acts for most applications, as a water-resistant barrier between the outside and inside of the building. The ability to resist the elements of the weather in penetrating the system is an essential quality to prevent moisture from collecting within any component of that system.

The essential elements of providing this weather resistant barrier depends on the following:-

1.   Quality of materials

2.   Design of the installation

3.   The installation being installed correctly

4.   Good workmanship

The basecoat layer of render, as applied to the insulation layer, is the primary weather-resistant layer together with good design and perimeter edge sealants.

Surface finishes, whilst usually of a decorative nature are also important in the overall performance of the render and any cracks, degradation, efflorescence, blistering or de-lamination should be avoided and repaired as necessary. Failure of any protection layer can result in the loss of insulation value, stained interior finishes, corrosion of component elements or deterioration of the basic building structure.

All system joints i.e. expansion, control and aesthetic must be water/weather-tight, a continuous barrier without cracks or breaks.

Traditional building finishes in the UK have been used primarily and historically as a resistance against the elements, for example, heavyweight "dry-dash" finishes used progressively as altitude exposure rating and local weather conditions directed.

The use of "organic" mineral finishes such as local aggregates as a dashing stone was traditionally very effective and still remains so. A medium to heavily textured finish effectively breaks-up the concentration of driving rain and succeeds in spreading the exposed load, limiting or reducing potential problems due to water penetration.

Generally, finishes in the UK form a geographic pattern, historically going from the traditional "smooth" finishes in the far south to the most heavily textured in the far north reflecting the prevailing weather patterns south to north.

References   1   *Reinforcing Layers*   *Section Three - Item 3.2*
              2   *Render Design*   *Section Four - Item 4.5*

2.8   Durability

The system and its components should exhibit chemical and physical stability and resistivity over its expected life under normal conditions. The normally accepted standard of durability of the system, within the UK, is 25-30 years. To achieve this, the system and its components should be fit for its purpose and tested to exhibit chemical and physical stability, usually by the British Board of Agrément and/or BRE Certification Limited.

Individual components together with the entire composite assembly must perform adequately to achieve the desired specified performance over the expected life of the system. The measure of efficiency is the level of the cost in maintaining the system throughout it's estimated life.

Under normal conditions the composite system should resist deterioration due to the action of the weather, environmental contamination, insects, fungi or rodents. All material parts should be compatible with all other building materials with which they come in contact. Incompatibility can lead to premature failure such as unprotected aluminium in contact with cement.

The application of individual components should be installed in accordance with the system suppliers instructions as any unauthorised deviation from the specific requirements of the supplier or designer may result in unspecified performances and invalidate any guarantee.

The end-user should ensure good planned maintenance checks are carried out on the completed external wall insulation system. This may be simply a visual check annually or bi-annually as any householder would carry out on any element of his or her property. Subsequent visual checks can pay dividends in keeping future costs down to a minimum or even eliminating any un-necessary future expenses.

When a system is being installed, initially as with every aspect of building construction, attention to good detailing is important for long term durability but in addition, the system provider should supply the end-user with a maintenance manual and guide.

Problems with external wall insulation systems arise post contract when building owners take a blinkered view of general maintenance and "good house-keeping" by not carrying-out simple preventative measures. Maintenance is a simple and straightforward task, which like many other measures in buildings are necessary on a regular periodic basis.

Building Owners should not alter, change or interfere with the system without consultation with the system supplier, contractor or designer.

References   1   Maintenance Manual   Section Six - Item 6.15

2.9   *Appearance*

The finish coat should be able to resist all normal exposures expected from the location, height, orientation and design of the building.

Resistence to erosion from weather, water penetration, solar radiation, wind, and temperature fluctuations are required to maintain good appearance. The system finish coat forms the visible exterior face of the project and its acceptability by the Building Owner will depend on the quality of materials, costs, installation and workmanship offered by the installing contractor.

The final appearance should be tasteful, attractive and pleasing to the eye with consistency whatever the finish. Flatness tolerances should be maintained throughout to ensure a consistent system appearance.

Local Planning Regulations may dictate colours and types of finishes to maintain the design aspects of adjoining buildings etc..

Architectural features may also play a significant part in the appearance of a final design and installation. Such features may demand detailing requiring special or unusual components to enable these features to be adequately and correctly constructed.

References   1   Photographs   Dashes
Scratch Plaster
Roughcasts
Textured Coatings
Tile Hanging
Brick Slips
Terracotta Tiles

2.10 Detailing

Detailing architectural features on external wall insulation must consider that weather tightness is the primary consideration and all special forms, trims and accessories must achieve this property.

The appearance of the requisite trim must be considered on its merits to provide an acceptable finish.

Examples of required detailing are as follows :-

Insulated areas below ground level
Base trims at low levels
Mid trims at floor levels
Fire breaks - vertical
Fire breaks - horizontal
Window and door reveals - Insulated
Window and door reveals - Uninsulated
Window overcills
Window undercills
Eaves and verges
Corner quoins
Raised feature window bands
Raised feature door lintels
Window pods
Ashlar Cuts
Bandings - Recessed
Bandings - Raised
Overhangs
Avoidance of "Cold-bridging"
Changes in wall thicknesses - External Faces
Changes in finish materials
Changes in finish colours
Changes in insulation, type and thicknesses
Changes in wall construction
Copings and cappings

Attachments to the substrate through the insulation system
Service ducts and covers
Attached external gates and fences

Detailing requires the appropriate materials manufactured to the correct shape and profile to ensure durability and be weather-proof.

Stainless steel trims and beads with powder-coated aluminium architectural feature trims are preferred, galvanised products which are unprotected are to be used with care if used externally.

The selection and suitability of particular detailing will be affected by the irregularity of the substrate and methods used to remove such irregularities.

References   1   Trim references   Section   Three   - Items 3.11 etc.

2.11 *Mastics & Seals*

Mastic seals used to prevent the ingress of water into the system must be sufficient in quality and application to perform adequately for the expected life of the system. Sealant joints are an integral part of any external wall insulation system and are essential in providing resistance to the ingress of water from whatever direction or source.

As external wall insulation systems require the full benefit of sealants to ensure complete protection of the system, full consideration to quality and workmanship in installing such sealants is essential.

The most common types of sealants used in the application for external wall insulation use are low and medium modulus multi-component polyurethanes and single component silicones.

Careful attention must be given to proper sealant selection and placement. Quality workmanship and an adherence to high standards of application are also very important as great reliance will be placed on a workmanlike installation.

The installation of all sealants is preferred to be carried out by the external wall insulation system contractor to avoid split responsibilities in the unfortunate event of any failure. An important characteristic of any sealant is the movement capability, the amount of joint expansion and compression a sealant can sustain without failure.

The selection of sealant will be determined by the joint size and width, calculation is based on the expected joint movement and the specified sealant movement capability. Allow for construction tolerances.

References   1   Cill Details   Section Four - Item 4.6 etc.

2.12 Avoidance of Risk

In selecting the appropriate external wall insulation system on any particular project there will always be differing requirements to ensure long-term durability and the minimum of maintenance.

The following are designed to high-light the principal considerations in the selection of an external wall insulation system to avoid un-necessary risks.

Selection of the correct system for the project

Certification - Use only a certified external wall insulation system as certified by the British Board of Agrément (BBA) or BRE Certification Limited. (Wimlas)

Exposure to use - Carefully select a certified system that will adequately survive uses associated with the building, it's location and exposure to weather, vandalism and any other consideration that may be imposed on the system to maintain it's appearance and durability.

Exposure to weather - Check the exposure to all weather conditions, location to exposures such as coastal exposures and heights above sea levels and orientations.

Compliance with local Planning Regulations - Check with the Local Planning Authority as to any detailing or finishes required to comply with any locally enforceable regulation such as Conservation Areas, National Parks and any other authority such as Land Owners and Non-Governmental Departments.

Fire Spread due to combustible insulation Selection of insulation material should consider the height of the building to be insulated. The more economic insulants such as expanded polystyrene are used successfully at low levels whereas the "firesafe" insulation materials should be used on the higher multi-storey buildings.

Interstitial condensation - calculations should be completed prior to any project being commenced to ensure interstitial condensation is avoided. This may involve varying the insulation thickness, vapour permeability or finish type to suit the original structure.

Internal surface condensation - calculations should be completed prior to any project being commenced to ensure internal surface condensation is avoided. Particular attention should be made to the use of the rooms to be insulated as high vapour-bearing internal atmospheres can contribute to the formation of condensation on internal cold surfaces.

Ventilation - ventilation in the form of trickle-vents or heat recovery systems will reduce the humidity levels of the internal air and thus reduce the risk of internal condensation. Careful consideration must be given to analyse every aspect of ventilation or if any, heat recovery with calculations, so that the desired results are obtained.

Impact Damage to System Careful consideration should be given to the locality of the cladding to ensure a sufficiently durable system is specified. Particular consideration should be given to pedestrian closeness which can lead to vandalism. Once vandals know that the system can be attacked, severe damage can be anticipated. Under these circumstances a heavy-weight system reinforced with stainless steel mesh is usually appropriate.

Condensation at thermal Bridges Condensation is a particular nuisance where openings in existing structures are continuous forming a "cold-bridge". This situation can be particularly difficult to insulate if dimensional constraints forbid the use of adequate insulation to cure these unwanted effects. Economies are sometimes prone to be contributory if sufficient insulation cannot be used in the right/wrong places, such as restricted window and door reveals.

Drainage to system   Due to severe exposure to wet weather conditions, thought must be given to shedding water off the system.   Failure to consider how this water can be drained away from the building can contribute to water-retention in the wrong places causing deterioration of the system long-term.

Seals and detailing   Detailing of the external wall insulation system to prevent the ingress of water is essential to avoid degradation of the insulation system.   The provision of the correct design and manufacture of trims and beads is essential in securing water-tightness.

Finish for project   The correct finish for the external wall insulation system is essential to :-

1.   Comply with planning requirements

2.   Be visually acceptable and be sympathetic with adjoining properties

3.   Be durable and fit for its purpose for its location and exposure

4.   Withstand the expected use such as normal wear and tear.

References   1   *Condensation*   *Section Four - Items 4.1 etc*

Part Two - System Performance

PART THREE - Materials and Components

3   The typical external wall Insulation system comprises many components to enable it to insulate the building, be secure and waterproof and pleasing in appearance.   The materials used and specified are required to be adequate and fit for their purposes, tested where appropriate and available to contractors in reasonable sizes and quantities.

Consideration should also be given, when choosing types of insulation boards, to the energy expended in the manufacture of such insulation and the ultimate benefit of energy savings as a result of its use. Individual manufacturers may be able to provide additional information in this respect.

3.1   Insulation Slabs

The insulation material supply industry, within the UK, provides a variety of rigid slab insulants suitable for the inclusion in external wall Insulation systems.

They are:-

Mineral Fibre (Rock and Glass based)
Expanded Polystyrene
Extruded Polystyrene
Polyisocynurate (PIR)
Phenolic Foam
Foamed glass

The above insulation types are usually supplied in rigid slab form to a variety of thicknesses and densities to provide the insulation needed for the particular project and to provide the specifier with the standard of insulation required by the client, building owner and/or Building Regulations.

Conventional slab sizes suitable for site installation are usually 1200mm x 600mm but some mineral fibre slabs are provided 900mm x 600mm & 1000mm x 500mm.

Mineral fibre slabs   Mineral fibre insulation slabs are manufactured from molten glass or rock which is spun and then pressed lightly into a slab at a density of between 80 and 140 kg/m³. Resins and binders are added to give the slab better resistance to water and to aid rigidity.

Mineral fibres are not good insulators on their own and obtain their insulation properties by entrapping air. If the material is compressed the air content will be reduced, or if some air is replaced by moisture in even small amounts, the thermal resistance can be reduced dramatically. Dry slab mineral fibres in the density range of 80-140kg/m³ have a thermal conductivity of around 0.038W/mK. Compression or the presence of moisture in mineral fibres should be avoided wherever possible, when applied to external walls.

Even though the industry makes a great deal of use of mineral fibre, it would appear to have its greatest advantage in that it provides good resistance to burning in the event of fire. Generally classed as "Fire-Safe" the rock based product has historically been specified on multi-storey buildings which require less consideration in the detailing of "fire-breaks". Most systems usually use these slabs at a density of 80-120kg/m³, offering a good degree of work-ability in being reasonably rigid and resistant to surface decompression.

Rigid mineral fibre slab possesses superb fire properties and are rated non combustible when tested in accordance with BS 476: Part 4: 1970 (1984). They also achieve Class 0 as defined in the Building Regulations.

The melting point of mineral fibre is in excess of 1000°C.

The thermal conductivity of mineral fibre   at a mean temperature of 10°C is in the range of 0.033 to 0.039 W/mK, depending on the grade.

Mineral fibre slabs repels water due to the presence of water-repellent additives. Moisture condensing from the air within mineral fibre is less than 0.02% by volume at 95% relative humidity.

When calculating vapour diffusion through a structure, the vapour resistivity of mineral fibre is negligible, resulting in reducing the risk of condensation and allow natural drying out of the construction due to it's ability to 'breathe'.

Mineral fibre slabs are easy to handle and install. It is easily cut to shape or size with a saw or sharp knife.

The slabs are usually supplied in shrink wrapped polyethylene packs, which provide short term protection. For long term protection, the slabs should be stored safely in a building. If it has to be stored outside, it should be stacked clear of the ground and covered with a securely anchored polythene sheet. Mineral fibre slabs should not be left exposed to the weather, if it does become saturated, it should be allowed to dry out naturally before the finishes are used.

No CFCs or HCFCs are used in the manufacture of mineral fibre slabs.

Expanded Polystyrene Raw bead polystyrene consists of spherical particles containing an expanding agent, pentane. It is primarily intended for the production of expanded polystyrene insulation slabs of a variety of thicknesses.

Bead polystyrene is supplied as a fire retardant moulding grade raw material used mainly for the production of block mouldings suitable for cutting into slabs. It is also suitable for the production of tiles, coving and other decorative mouldings.

Bead polystyrene (FRA Grade) has good fire retardant properties and will meet the requirements of many national standards including DIN 4102 Class B1 and B2, BS 3837 (1977) and BS 3932.

Bead polystyrene is processed in two stages; pre-expansion and moulding. When heated above 90°C, usually by direct steam, the blowing agent dissolved in the beads vaporises and the pressure so generated causes the softened polystyrene beads to expand to occupy 20-40 times their original volume. The resultant pre-foam consists of separate enlarged beads each having a large number of non-interconnecting cells.

After a 'maturing' period, generally 5-48 hours, during which air permeates back into the expanded beads, the pre-foam can be moulded, usually by direct steam, to the required final shape or block.

The most popular thermoplastic insulation slab is expanded polystyrene bead board. This comes in two types, virgin bead and recovered bead. Virgin bead boards will be less likely to take up moisture since each small discrete sphere of polystyrene has a skin round the outside which offers a degree of protection against water: in recovered bead board, this skin is destroyed. The thermal conductivity of virgin polystyrene bead board is generally taken to be 0.038/mK and since it is resistant to the uptake of moisture this may be assumed to be constant throughout.

In addition, the cellular nature of this material means that although it is resistant to water in its liquid sate, it still allows moisture vapour to pass through - an essential property in this application to the exterior insulation of buildings. Other desirable characteristics are that it is easily handled, behaves well in adverse conditions and is commonly available in most parts of the world. Generally economical in use at present levels.

Available standards are as follows:-

EHD Extra High Duty
HD -High Duty
SD -Standard Duty

The ultra-violet component of sunlight causes slight degradation of polystyrene which shows as a marginal surface yellowing. If the surface is exposed to wind and rain the degraded layers are very slowly eroded.

Expanded polystyrene has a low thermal conductivity, the value of which depends on its density and the temperature.

Typical values are determined by BS 874 1956 Appendix A Method 1 at 10°C.

Density kg/m3	W/mK
16	0.035
20	0.033
24	0.032

Expanded Polystyrene (EPS) is available in two "Types" -

1.   Flame Retardant Additive (FRA)
2.   Normal (N).

Flame Retardant Additive expanded polystyrene contains a sufficient concentration of a suitable additive, or is otherwise modified, to ensure that when tested in accordance with the method in BS 4735: 1974, shall show an extent of burn less than 125mm.

However, this should not be construed as a means of assessing the potential fire hazard of the material in use.

The British Standard Specification BS 3837 for Expanded Polystyrene (EPS) board which calls for products to comply with set physical properties and both require all board to be colour coded by Grade depending on the use for which they are required. as follows:-

Classification Identification
Extra High Duty   Green stripe
High Duty   Black stripe
Standard Duty   Yellow stripe

General Guidance Notes

1. EPS should be stored away from inflammable materials.

2. Storage and working areas should be kept free from rubbish.

3. All personnel must know that EPS is combustible.

4. Work which requires flames or burning should not be carried out without fire extinguishers being to hand.

5. Unauthorised access to the storage and working areas should be restricted to reduce the risk of fires being started by accident or intention.

6. If there is an outbreak of fire, the fire brigade should be called immediately.

Note :- Expanded polystyrene is CFC and HCFC free.

Extruded Polystyrene Extruded polystyrene, a closed-cell foam, has several additional merits over expanded polystyrene as it is the least prone of all the more economic semi rigid insulations to the uptake of moisture. Such foams do not absorb water or become waterlogged even under severe conditions thus avoiding any loss of thermal performance.

The thermal conductivity value for extruded polystyrene, at approx 0.027W/mK for the best quality.

The ability to inhibit the passage of water either in its liquid or vapour states may inhibit the breathing performance of the building.

All polystyrenes are thermo-plastic materials which means they exhibit plastic characteristics when subjected to temperature rise. With a semi rigid board, however, some stresses may be expected to occur at the interface of this material and the rigid coating.

The most useful area to use this extruded polystyrene board is below any damp proof course, or below ground level, where there is a possibility of moisture penetration into the insulation layer.

Available in many board sizes but 1200mm x 600mm is the most suitable. A wide variety of thicknesses (Fire Retardant grade) are available.

Polyurethane & polyisocyanurate Polyurethane and their cousins, the fire retardant (but more smoke producing) polyisocyanurates have almost the lowest thermal conductivity of insulating materials in common use. While their initial thermal conductivity value is definitely in the "super-insulant" class, being in the order of 0.022-0.025 W/mK.

These materials may be more susceptible to moisture uptake than any of those mentioned previously, (with the possible exception of mineral fibre).

To reduce the likelihood of moisture ingress the manufacturers usually offer their board products as a sandwich of foam between two layers of paper. These papers incorporate bitumen, polythene, glass tissue or aluminium foils. While these work very well on the front and rear surfaces of the board, the edges are not protected, and polythene, kraft paper and foils present very difficult surfaces for fixing and further finishing, and may inhibit the vapour permeability of the wall.

Aluminium foil-faced boards are to be avoided if using cementitious finishes as the alkalinity of the render will attack aluminium foil. Glass tissue faced boards may be used for external wall insulation systems.

Some polyurethanes can be foamed in-situ but these are not suitable for external wall insulation when an even, flat surface is required.

Available in many board sizes but 1200mm x 600mm is the most common. A wide variety of thicknesses are available.

Phenolic Foam Phenolic or phenol formaldehyde foam is manufactured by mixing phenol formaldehyde resin and a blowing agent and curing the resultant board. On emerging the board is sliced by passing it through a bank of saws and the board thus obtained is fairly brittle and friable which will not burn or give off heavy toxic gases in a fire.

The insulation board comprises a C F C free rigid phenolic insulation core with plain glass tissue facings on both sides manufactured in accordance with the requirements of BS EN ISO 9002: 1994.

The thermal conductivity value is about 0.019W/mK when dry. However, the cell structure of the material is such that it is water permeable and therefore manufacturers of these boards usually offer them with a foil, polythene or glass tissue covering back and front. This also helps to combat the friability of the material.

The outstanding advantage of a phenolic material is its behaviour in a fire, and whilst not classified as "fire safe", its resistance to fire is exceptional. Where users are concerned about fire, it is an option to be considered albeit fire breaks may still be required under certain circumstances.

Aluminium foil-faced boards are to be avoided if using cementitious finishes as the alkalinity of the render will attack aluminium foil. Glass tissue faced boards may be used for external wall insulation systems.

Available in many board sizes but 1200mm x 600mm is the most common. A wide variety of thicknesses are available.

Note :- All thermoplastic insulation boards should be allowed to mature immediately after manufacture to increase their effectiveness as insulators.

Manufacturers should be contacted for further information as to their specifications in respect to the classification of "fire-safe".

74

Foamed Glass   Foamed glass, when applied externally, can be used with a variety of final finishes such as render, tiles or profiled metal.   It can be applied directly to the external surface of the wall providing an opportunity for external upgrading. Totally impervious to water and water vapour, seepage of water through poor external wall construction is eliminated whilst insulation values of the whole wall are improved.

Foamed-glass board is an ideal rigid insulation material for use with external wall insulation systems as it can be applied directly to existing poor quality external wall surfaces. Its rigidity allows the board to be "packed-out" as necessary to straighten wall surfaces and will span small voids.

Foamed-glass does not deteriorate with age. ensuring that these values remain constant.   Foamed-glass is rigid and dimensionally stable. so it does not shrink or warp with the effects of climatic variance. With only a minute coefficient of expansion, (approximately that of concrete). there is no danger of joints between the boards opening up .

Foamed-glass is totally non-combustible and will not give off smoke or toxic fumes, it can be classed as "fire-safe".   It is completely free from CFC and HCFC .

Available in many board sizes but 1200mm x 600mm is the most common. A wide variety of thicknesses are available.   The thermal conductivity of foamed-glass insulation is approx 0.038W/mK.

Insulating Render   Wet applied insulated render is a low density cement/sand mix into which polystyrene beads are mixed. This is a matrix insulation system and its physical characteristics are similar to those of low density air entrained concrete of about the same weight.

Insulating render has a typical thermal conductivity value of approximately 0.10W/mK, a high comparative cost together with a density of 400kg/m³.

The main uses of this material is as follows :-

1.   Render and insulate small inaccessible areas such as window and door reveals, thus providing a reasonable surface and combining a degree of insulation to reduce possible effects of "cold-bridging".
2.   Insulating Listed Buildings or buildings in conservation areas where it can be installed internally, thus avoiding external changes of appearance.
3.   The light-weight characteristics of the system also has advantages in certain cases. (Average of 10.5Kg/m2)
4.   Insulating render is useful in areas of impact risk, whereby it can easily be repaired after frequent impacts or after frequent contacts due to abrasion.

3.2   Reinforcing Layers

An external wall insulation system has to include a tough over-coating to shield the insulation from weather, support the final finish and casual impacts and vandalism. These systems are reinforced by either using a metal or mineral fibre mesh.

Reinforcing layers for render coatings come in many forms, but the most convenient to use are expanded metal, plastic and glass meshes.

Methods of application are usually either :-

a.   As a metal, polymer, glass mesh or armature installed on the surface of the insulation board between the layers of coatings or

b.   As a mix of chopped fibres in the wet render, before or whilst it is being applied.

The ultimate choice is critical to the success of the whole external wall insulation system as there are a number of forces operating requiring careful selection, namely :-

a.   Impact resistance
b.   Movement control
c.   Durability

The choice of reinforcement layer, which will decide the final system selection, will be made by the Contract Supervisor, dependent on his requirements to satisfy local contract conditions.

Metal Reinforcements   Metal reinforcements are provided in a variety of materials namely:-

Austenitic Stainless Steel
Ferritic Stainless Steel
Galvanised Mild Steel

A description of these metals is as follows:-

Austenitic stainless steels are non-magnetic. When nickel is added to stainless steel in sufficient amounts the crystal structure changes to "austenite". The basic composition of austenitic stainless steels is 18% chromium and 8% nickel.   This enhances their corrosion resistance and modifies the structure from ferritic to austenitic. Austenitic grades are the most commonly used stainless steels accounting for more than 70% of production (type 304 is the most commonly specified grade).

Ferritic stainless steels are plain chromium stainless steels with a chromium content varying between 10.5 and 18% and a low carbon content.   They are magnetic and not hardenable by heat treatment. Ferritic alloys have good ductility and formability but a relatively poor high temperature strength compared to that of austenitic grades.

Galvanised steel is a mild steel product with a coating of zinc. The zinc protects the steel by providing cathodic protection to the exposed steel, so should the surface be damaged the zinc will corrode in preference to the steel. Galvanised steel is one of the most widely used products, used extensively in the building sector, automotive, agricultural and other areas where the steel needs to be protected from corrosion.

Metal reinforcements or armatures (not aluminium which is attacked by the alkaline cement) can be used in a variety of different forms such as welded wire, expanded metal mesh or as steel fibres chopped into short lengths and mixed into the render. Initially, mild steel would seem a reasonable choice since it is well known, very strong, and is commonly used as a reinforcing medium for concrete. However, in reinforced concrete, if rusting is to be prevented, the steel is required to have at least 50mm of concrete cover, but in a 25mm thick render, this is not possible. Any steel used must therefore, be galvanised coated or stainless steel.

The choice between the two is dictated by the thermal expansion coefficient of the metal. Normally in reinforced concrete, thermal expansion is not a major problem. However, within a thin layer of render and with insulation immediately behind the metal, any thermal shock or temperature build up will quickly affect the metal. Since there is no depth of masonry here to absorb and dissipate the heat, the temperature rise may be expected to be substantially higher than that of conventional concrete.

Any attempt to incorporate reinforcing with integral rods or channels will increase the likelihood of cracks occurring along the line of the additional metal. Not only will the expansion of the metal be higher but its extra concentration will drastically increase the stresses, built-up in render systems, thereby considerably increasing the risk of cracking.

Welded wire mesh Formed on automatic welding/assembly machines, these meshes are formed with round bars usually of approximately 1mm diameter, and welded at all contact points on a square grid of approximately 25-50mm. Other grid and bar diameter sizes can be used dependent on individual requirements.

Available in stainless steel, galvanised mild steel and PVC coated galvanised mild steel. This mesh offers good tensile strength combined with excellent performance in fire conditions.

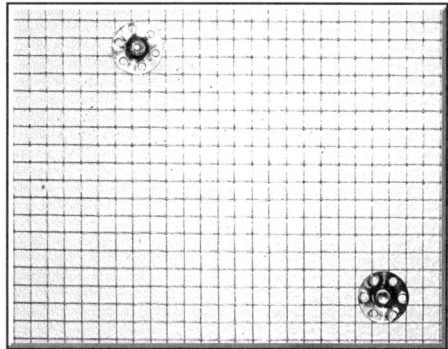

Grid-weld mesh

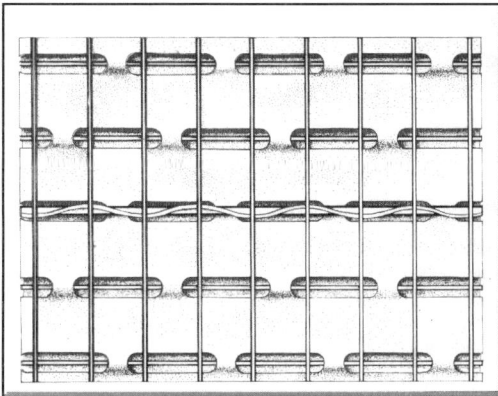

Courtesy of BRC Special Products

Some meshes are also available with a range of waterproof backing papers to help protect render and substrate.

One type of grid-weld mesh is available as a proprietary material having additional stainless steel wires interwoven with a layer of absorbent paper and backed by a water-resistant breather paper.

Expanded metal mesh   These meshes are formed by sheet steel being slit and expanded into a diamond or square pattern. The stainless steel versions offer good resistance to corrosion and its tensile strength will be relevant to its cross sectional area.

Riblath by Expamet

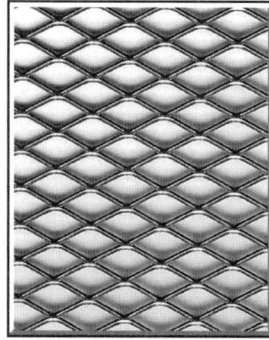

Exp metal mesh

Pre-galvanised expanded mesh has a low resistance to corrosion but all types of expanded meshes (except stainless steel) are available galvanised after manufacture which offers improved corrosion resistance.

All types of metal meshes offer good performance in fire and can be mechanically fixed to achieve good resistance to wind suction loads providing the correct anchors are used.

Notes:-

1. Care must be taken when nesting ribbed metal lathing so that concentrations of metal are minimised.

2. Where a fixing is to be used in conjunction with a metal mesh it is recommended that it is positioned at the junction or overlap of wires to maintain high pull-through resistance.

3. Care to be taken when selecting meshes as to exposure together with interstitial condensation.

Polymers and polymer meshes   The polymer materials normally adopted are nylon, polypropylene and polyester. None of these offer a perfect answer in itself, and in the long term a combination may be the solution.

Nylon is extensible and difficult for the matrix to bond to, thus creating problems due to lack of adhesion and binding.

Polypropylene, although it has been used as a random fibre in a cement mix, is difficult to bond to. Corrosion resistance is good and mechanical fixing is easily achieved.

Polyester is not very resistant to alkalis and as render is alkaline it cannot be considered as a long term reinforcement material.

Glass Fibre Meshes   Glass or mineral fibre may be used as a render reinforcement or crack preventing material, it is obtainable in three types namely :-

a.   Sodium Silicate or "soft" glass, which is not resistant to alkali attack and therefore cannot be recommended for this situation

b.   Borosilicate or "hard" glass which has a greater resistance to alkalis than soft gall, but which at temperatures in excess of 77°C (and in the presence of moisture) will be attacked. The alkali resistance may be improved by over coating - usually with a polymer layer. (SBR)

c.   Zirconium or AR glass which is highly alkali resistant and therefore will not be attacked in the life span of these systems.

Mesh types a. & b. are currently used as reinforcements and therefore merit consideration. While obviously Zirconium glass is the best choice, the rights to this material in the UK are patent protected and therefore not available to every glass manufacturer.

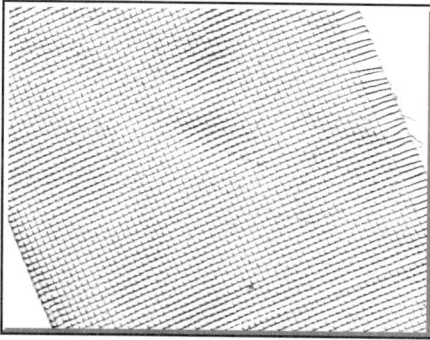

Typical mineral fibre mesh

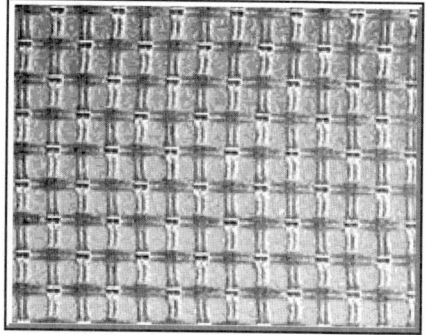

Mesh showing interweaving

Borosilicate, on the other hand, is no longer protected by patents and several glass manufacturers particularly in Eastern Europe use this material extensively.

It is very difficult to determine which grade of material has been supplied by merely looking at it and assurance must always be sought on the source and nature of any material provided.

Glass Mesh Fabric is a woven mineral fibre scrim available in various weights with the following properties :-

Alkali Resistant
Available in Standard, Medium. and Heavy Weights
Improves Crack Resistance
Improves Impact Resistance
Reinforces Stress Points
Reduces the need for Control Joints

Detail Mesh

Technical Specifications
Designation   Fibreglass scrim
Weave   Leno
Coating   Modified SBR (Alkaline resistant)
Self-extinguishing   No
Weight   100 - 200gm/m^2 (Light, Medium &
                                                  Heavyweight)

Aperture size   10 – 15mm
Rolls   50 x 1m
Shelf-life   Indefinite.

Fibre Matrices   Chopped fibres incorporated in the render require the addition of liquid polymers to achieve a suitable bond together with designed strengths. The external wall insulation systems incorporating these fibres require extra large headed fasteners situated at precise dimensions to provide a maximum span for the render between fasteners. The systems rely on exceptional quality control both for the fixing layout and mix quality as the design is critical to these factors.

3.3   *Fixings & Fasteners*

The use of mechanical fixings or fasteners to secure an external wall insulation system to the outside of any wall is well developed with a wide variety of fasteners available. These fasteners are supplied in three basic forms namely:-

A.   All plastic
B.   Combination of plastic & metal
C.   All metal

The design of fixing patten is governed by the weight of the system, its exposure, the quality of the substrate to support it and any other factor necessary to be considered for the safe installation peculiar to any particular project.

Determination of fixing design and performance is resulting from a "Pull-Out" test utilising a specialist pull metre.

Fixing manufacturers also provide a service to determine fixings performance on sites.

Weight of system

The total weight of the system must include the following elements:-

> Insulation slabs
> Trims, flashings and beads
> Fixings
> Render reinforcements
> Renders
> Finishes

Wind loadings

Generally the Code of Practice CP3 is the current guide as to the evaluation of wind loadings in relation to exposure to the wind forces within the UK and should be referred to for full explanation.

Fixing Types and Uses

The majority of which are the following :-

a. Moulded plastic fixings are hammered into pre-drilled holes, the fixing securing itself with the friction action along the embedded shaft length.

b. Self-drilling all metal fasteners are used for fixing to board sheathing

c. Spring-loaded locking fasteners are used for captivating cavities.

d. Concrete threading fasteners which thread themselves into a pre-drilled hole. (Including washers)

e. Screw fixings for timber supports (Including washers)

f. Coach screws/bolts for timber supports (Including washers)

Plastic Fasteners Usually supplied in polypropylene/nylon, these fasteners form the majority of used fixings as cost is generally a major consideration. They are hammered into pre-drilled holes usually 8-10mm diameter.

Plastic - Metal Fasteners
Generally supplied as the plastic fastener but with a metal centre expansion pin. The pin being manufactured in carbon or stainless steel.

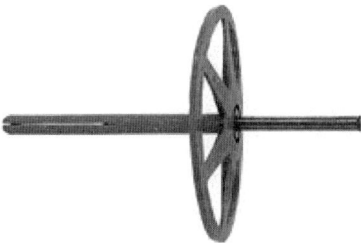

Fischer Fixings Illustrated

Metal Fasteners   Usually supplied in stainless steel, but can be supplied in galvanised mild-steel, these fasteners are generally specified where fire precautions are particularly prevalent such as multi-storey buildings over certain heights.

These fixings are hammered into a pre-drilled undersized hole where the shaft of the fixing is squeezed smaller to gain the greatest friction grip.

85

Concern has been expressed by some specifiers in respect to cold-bridging through the insulant by the use of these fixings, this has not been proven.

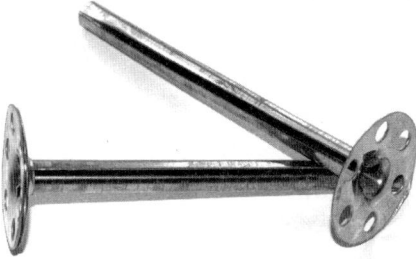

Ewinail 110

The illustrated fixing above is a stainless steel insulation anchor for use in the construction of fire breaks and where fire resistance assurance is demanded, i.e. where an additional stainless steel fixing, every square metre, is required. Usually available in four lengths, 50mm, 90mm, 110mm & 140mm, shaft diameter of 8mm, these provide for most conventional thicknesses of systems.

Extra large heads are also available with some fixings.

There are additional fixings available which are 200mm in length but with a diameter of 10mm.

The principal of the fixing in resisting pull-out loads is that the shaft of the fixing is oversized as to the drilled hole. The shaft of the fixing squeezes smaller as it is hammered into the hole and grips the side wall.

On site individual tests are required to establish the correct loadings.

Self-drilling Screws   A variety of hexagon-headed screws, supplied with a suitable washer-plate can be used to fix systems to timber cladding, cementitious board sheathing or metal sheeting.

Stainless Steel Washers   Stainless steel washer heads of various sizes are used to secure the insulation.

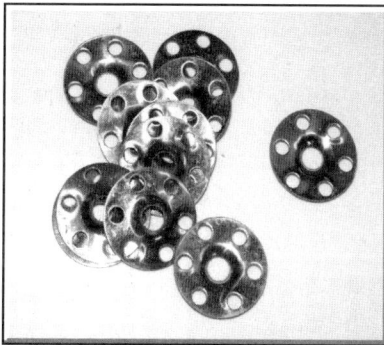

38mm dia washers

3.4   Adhesives

Adhesives for fixing insulation boards to a wide variety of substrates can be supplied in various forms, acrylic or cementitious. Usually supplied pre-mixed in buckets these adhesives are ready to use under normal working conditions and in accordance with manufacturers instructions.

All parameters for mechanical fixing design to be considered and complied with using adhesives.

Acrylic adhesives   Supplied as 100% acrylic polymer-based non-cementitious adhesive or base coat, to adhere rigid slab insulation boards to approved masonry sub-straits. When used as a base coat, a reinforcement layer of glass mesh is required.

Note :- There are some proprietary adhesives available which do require the addition of portland cement.

Polymer Cementitious adhesives   Supplied in liquid form, Styrene Butyl Resin (SBR), the polymer is added on site to a strong sand/cement mix to adhere rigid insulation slab to masonry substrates. When used as a base coat, a reinforcement layer of glass mesh is required. Dry powder mixed adhesives are proprietary prepared cement based adhesives, which only require the addition of water.

Dry powder mixes should be added to the clean water in a clean container and mixed thoroughly to give a fairly thick creamy slump free mortar of the desired consistency using a mechanical stirrer.

3.5   Renders, Coatings and Finishes

Cementitious renders and rough-casts can be divided into three types, heavyweight, medium weight and light-weight renders, reinforced as required with either stainless or galvanised mild steel mesh for heavyweight renders and mineral fibre meshes for medium weight and light-weight renders. All types of render will require a reinforced base coat and a finishing coat, the final finish being sometimes an additional further application.

All materials for the application of renders shall comply with the relevant BS Code as follows:-

Cements	BS 12, BS146, BS4027, BS 5224
Limes	BS 890
Sands	BS 1199
Aggregates	BS 882

Pigments	BS 1014
Plasticizers	BS 4887
Water	BS 3148
Expanded Metal Mesh	BS 1369
Welded Wire Mesh	BS 729
Stainless Steel	BS 1449

The majority of external finishes to building substrates on insulation materials are cement based renders, namely :-

a. Traditional wet applied cementitious base & topcoats
b Polymer modified cementitious base & topcoats

In addition

Pre-mixed, bucket supplied proprietary acrylics & silicones

These predominately form conventional thick-coat dry-dashed, rough casts, scratched or polymer modified medium coat coverings or lightweight thin coat textured coatings.

Cementitious Base Coats All these base coats are cement based and/or polymer modified, supplied pre-blended and pre-bagged, requiring only the addition of water and thorough mixing in accordance with manufacturers instructions.

The basecoat is reinforced with a metal or mineral fibre reinforcement layer.

Applied normally in one coat as a backing and lightly scratched finish, ready to receive further cement based finishes or as a trowelled finish for other commercially available proprietary finishes/coatings.

Heavyweight Basecoat is for general use as an 8-12mm base coat for cement based finishes. The mix will follow conventional uses of render, usually 4/1 sand/cement with waterproof additives and plasticisers. Applied with traditional tools in the traditional manner. Fixed directly through the mesh into the substrate.

Medium-weight Basecoat applied as above but generally 6-8mm thickness and reinforced with a mineral fibre mesh. Polymers and water-proofers added to improve overall performance. Can be adhesive or mechanically fixed.

Lightweight Basecoats applied 4-6mm thickness are specialised high polymer mortars reinforced with mineral fibre mesh. Can be adhesive or mechanically fixed.

3.6   Finish Topcoats

All systems using reinforced basecoats require to be finished with any one of the following types of topcoat namely :-

1.   Aggregate dash
2.   Scratch Plaster or Scraped Finish
3.   Roughcast Render
4.   Spray Render
5.   Coloured Smooth Render
6.   Paint Finish on Smooth Render
7.   Silicone Additives
8.   Textured Coatings

The basecoats are applied as specified with a lightly scratched finish. The topcoats are described as follows

Aggregate Dash   Dash Receivers are cement based, polymer modified and self coloured, requiring only the addition of water and a short mixing time. Normally applied as an excellent background for any dry dash or roughcast finish for new build or refurbishment.

All Dash Receivers should comply with ISO 9001 approved and conform to BS5262 Type 3. For general use, normally applied onto proprietary basecoats.

Apply a butter coat of render to a uniform thickness 6-8mm, depending on the aggregate size. Whilst the render is still plastic, throw washed aggregate on to the surface to give a uniform dense coverage. Immediately tamp the aggregate particles lightly into the butter coat with a wood float and ensure a good bond is obtained.

The greater the size of aggregate the thicker the dash receiver to suit.

Scratch Plaster or Scraped Finish   Scratch Plaster is a pigmented specialist mortar with a large grain size that will provide a textured finish when scratched. Supplied pre-mixed and pre-bagged it only requires the addition of water. Applied onto a standard basecoat to a thickness of 10-12mm, the final surface is scraped with a scratching tool to remove 2-3mm of material in order to achieve the correct desired texture.

To achieve a textured finish, scraping should take place when the render has set but not hardened. The exact timing of this operation will vary according to the weather conditions.

The surface is ready when a thumb impression cannot be made and when the aggregate scrapes easily from the matrix without sticking to the scraper.

Scraping should be carried out in a tight circular motion, and the surface brushed down with a soft brush upon completion. All areas must be scraped at the same stage of readiness, as early scraping will result in darker shades and late scraping in lighter shades. A uniform approach is essential to achieve an even finish.

Brush-down on completion with a stiff brush to remove surplus material.

Roughcast Render   Standard Roughcast is a cement based, polymer modified, self-coloured render with aggregates to provide a self-textured finish. It requires only the addition of water and a short mixing time.   It is applied usually by hand-throwing onto a coloured basecoat which provides an excellent low maintenance alternative to dry dash and traditional painted finishes.

Standard Roughcast should comply with ISO 9001, and should conform to BS 5262 Type 2.   For general use, normally applied to Standard Dash Receiver.

First, apply a Standard Base Coat as an 8 -10mm primary coat over block-work.

Second, apply a butter-coat of Dash Receiver to a uniform thickness of approximately 4-8mm.

Mix the Roughcast in accordance with the manufacturers instructions to obtain a wet plastic mix.   While the Dash Receiver coat is still green, throw the Roughcast on to the surface, using a hand scoop or a mechanical applicator.   Great care should be taken to obtain a wide, even spread.

Spray Render   Spray render mixture is applied, by hand, in light sprays from a spatter machine, passing in an arc from 45° left to 45° right, and returning over the same area.   Each pass will apply approximately 2kgs per square metre, with a thickness of 1.5mm per pass.   A minimum coating thickness of 4mm with a weight of 5kgs per square metre is usually necessary to achieve a good density of coverage.

Coloured Smooth Render   Pre-mixed and pre-packed renders are available in a variety of colours, particularly for scratch plasters and dash receivers, the following are a selection currently available.

Selection of Aggregate Finishes

Black & White Aggregate

Derbyshire Spar

Arcanc

Orchid

Pink Champagne

Canterbury Spar

Yellow Aggregate

Textured Coatings

Bark Effect

Fine Finish

Textured Finish

Silicone Effect

Silicone Effect with Render

Selection of Projects

Cornish Houses

Brick Slips & Render

Texture Coating

Classic Effects

Multi-Storey Flats

Mixed Finishes

Selection of Details

Brick Slip Detail

Terra Cotta Tiles

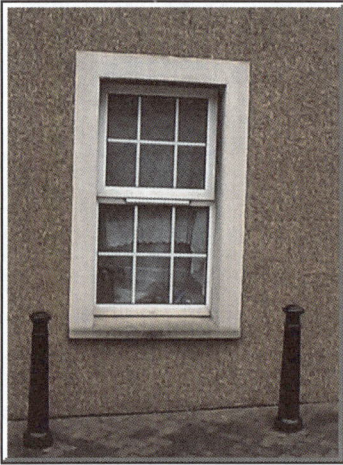

Window Bands

Brick Render

Tile Hanging

Dash Effects

Painted Smooth Render   All paint finishes are to be applied in accordance with the manufacturers instructions. Preparation of the sub-strate is essential to ensure a smooth and even surface ready to receive the paint application in however many layers or coats.

Silicone Additives   Renders are now available with silicone additives, specially designed to incorporate the benefits offered by these water repellents.

Silicone adds a high water repellent quality, whilst allowing water vapour to pass freely through the render, thus the amount of dirt adhering to the surface is greatly reduced ensuring a freshly rendered appearance for a prolonged period of time.   This dry surface also improves the resistance of the finished render to algae growth and the natural phenomenon of lime-bloom.

Textured Coatings   Supplied in buckets usually of 15kg weight, this product is applied over a prepared base to suit requirements, either lightweight or heavyweight.

Acrylic or silicone based, these coatings are applied by trowel or roller to an approximate thickness of 1.5 to 3mm dependent on aggregate size.

Rollers can determine differing finishing effects.

3.7   *Natural Hydraulic Lime Renders*

The knowledge and experience of manufacturers who maintained the production of hydraulic lime throughout the 20th century, solely concentrate on the production of natural hydraulic lime. These products are produced in a modern processing plant, maintaining the principals of producing traditional natural materials which meet to days requirements for quality and consistent performance. They are pre-blended and pre-packed to ensure quality control and consistency.

Modern lime mortars will allow finishing to a considerable scale without the requirement for expansion joints when used as mortar, plaster or render. The requirement is for the replication of historical finishes only capable of being replicated by lime mortars and can be important if external wall insulation is desirable for good reason. The initial set of lime mortars allows application on damp or salt laden backgrounds without detriment and enables further construction to be undertaken and only requires approximately one day additional setting time compared to cementitious mortars, dependant on site conditions.

The preparation of the background should be thoroughly brushed to clean away salts and other loose deposits, whether new basecoat render or old sub-strate. Where the sub-strate is excessively dry it should be dampened down prior to application. Arrises should be formed traditionally using timber lathes avoiding the use of metal or plastic beads.   At the junction of dissimilar materials in the background, reinforce the substrate with mesh.

Finishing lime mortars can be applied, by metal trowel, in layers of 2mm thicknesses as required and finished smooth off the trowel and sponged to an even texture. External wall insulation can be rendered with lime mortars but extreme care is necessary to apply and finish the render, follow manufacturers instructions.

Brick slips can be pointed with lime mortars to replicate traditional lime joints as and when required.

3.8   *Brick Slips*

Brick slips are thin bricks either cut from full sized bricks or fired in the slip form by the brick manufacturers. The slips are usually 6mm, 15mm, 20mm or 25mm in thickness but 20mm thickness is the preferred thickness. Brick slip corners or "pistols" as they can be known are usually cut from full sized bricks as the firing of corner slips in the UK is difficult due to poor clay materials.

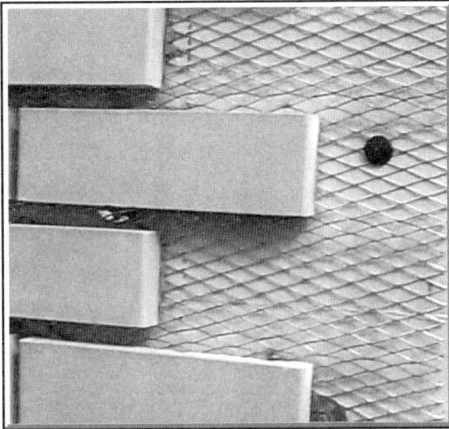

Brick slips on mesh

The layout of the brick slips is planned to provide a similar layout to that of real bricks to include all of the traditional popular bonds.

E.g.
Flemish bond,
English   Garden   Wall
Bond,
Stretcher Bond
            and many more.

The same considerations are given to the general overall appearance of the slips as for real bricks, that is;

a.   *Mixing the batches of slips to even-out colour variations*
b.   *Sorting for face texture and graining.*
c.   *Careful handling on site to avoid chips*

Extract from BS 3921 : 1985

The average size of the bricks in a batch should not vary greatly from the average size of 24 bricks tested, although average size may occasionally alter sufficiently from one batch to the next for the effect to be noticeable in the finished brickwork. Individual bricks, on the other hand, may show a greater deviation due to differences cancelling one another in the test. These variables are a natural characteristic of fired clay bricks. Accuracy can become critical where very short lengths of brickwork are involved and insufficient joints are available to absorb variability in the individual bricks without excessive variation in joint width. It is also important where, in a larger area of work, a poor appearance might result from a change in the average size (and hence joint width) due to differences between batches. In such cases the designer should consider carefully the size and character of the particular bricks specified.

At any one time, variations in the manufacturing, drying and firing conditions will cause sleight variations in the size of the bricks. The magnitude of variation in the batch is influenced by the type of clay and the manufacturing process.

In addition, the batch average can deviate from the planned brick size due to gradual changes in raw materials, for which the manufacturer has to make periodic adjustment.

Brick slip & render combination

Appearance

The appearance of brick is always a matter of agreement between the specifier or user and the manufacturer or supplier.

The visual requirements of the Architect will vary according to the type and colour of the bricks selected. This will depend on the overall design and the inherent characteristics of the bricks, e.g. common, facing, handmade or stock.

As a guide bricks should be reasonably free from deep or extensive cracks and from damage to edges and corners, from pebbles and from expansive particles of lime.

It is essential to build a sample panel for reference so that it reasonably represents the finished work and exposes for assessment those faces which will be visible in the finished work. In particular, bricks should be laid in the bond selected for the finished work, using mortar of the same colour.

The joints should be tooled in the same manner as the finished work.

NOTE:- A viewing distance of 3m is normally the standard for the purpose of the assessment. This may be varied by prior agreement between the supplier and the specifier.

Brick slip cladding systems onto insulation can be applied in a variety of ways, the essential principal being security of the slips for safety reasons. Brick slips falling from a great height are dangerous so particular consideration must be given to brick slip security so as to avoid any de-lamination from the insulation layer.

Project showing a range of finishes

After safety, all other considerations as for renders must be observed and complied with to ensure a satisfactory performance with durability and water resistance.

Profiled Polystyrene   A very popular method of fixing brick slips is with the use of profiled polystyrene.   The insulation is profiled into standard brick courses to carry and set-out the slips. These systems rely entirely on the adhesion by the brick slip to the surface of the polystyrene

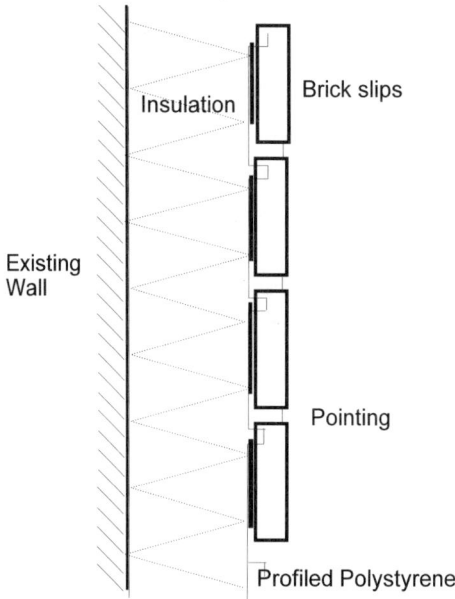

The expanded polystyrene is easily "hot-wire" cut into brick   courses,   to accommodate bricks   of the following dimensions :-

50mm
65mm
70mm

The brick slips are bonded into the cut courses with a good   proprietary   water-bound adhesive.   Solvent based adhesives cannot be used.

Brick slips can be extruded or cut to the following thicknesses :-

10-12mm
12-15mm
15-20mm

The pointing is a specialised waterproof and coloured pointing mortar, the joints being formed in the conventional manner. Recessed joints may only be used if the thicker slips are used and the remaining depth of pointing is suitable to be weather-proof and durable.

Profiled Stainless Steel Mesh
Featured Specialist System - EPSIBRICK

Variety in insulation types and thicknesses can be used and fastened to the substrate in the conventional manner.

The profiled brick mesh is layed over the insulation and fixed to the substrate

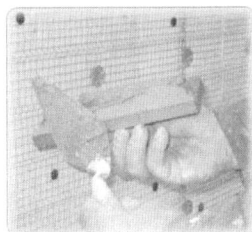

The brick slips are planned in accordance with conventional brick layouts and cut on machine as necessary

The brick slips are coated on the back face with adhesive mortar and pressed in position.

The brick slips are aligned vertically to ensure the slip edges are perpendicular.

The slips are pointed by gun or bag and tooled to a conventional finish.

The brick face is brushed free of surplus material on completion.

103

Laminated Skin-Insulation System
Featured Specialist System - EUROBRICK

Eurobrick Systems introduced to the UK market in 1991 uses standard sized bricks externally and internally, together with the insulation thickness, to achieve current thermal insulation standards.

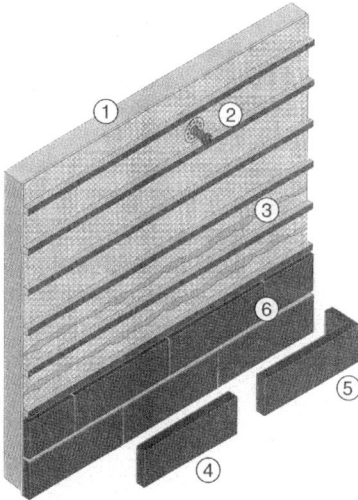

The system comprises the following :-

1 - Insulating backer panels

2 - Panel fasteners

3 - Permanent bonding adhesive

4 - Brick slips

5 - Corner bricks

6 - Dry-pack mortar

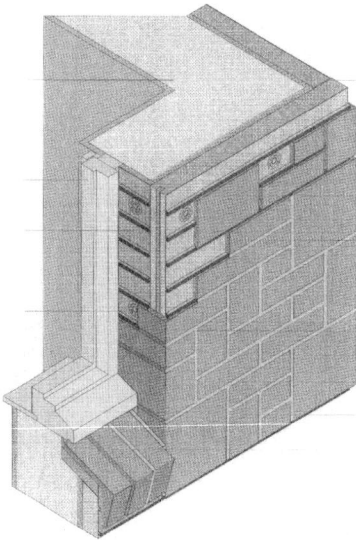

The insulating panels comprise of extruded polystyrene sheets laminated with a vacuum formed, profiled, waterproof skin. The ribs in the skin align to form brick tracks and give mechanical support to the slips. Insulation fasteners connect the sheets to the substrate. The brick slips being permanently bonded to the waterproof skin and pointed with the dry-pack pre-blended specialist pointing mortar.

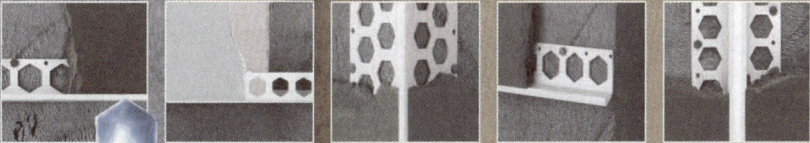
105

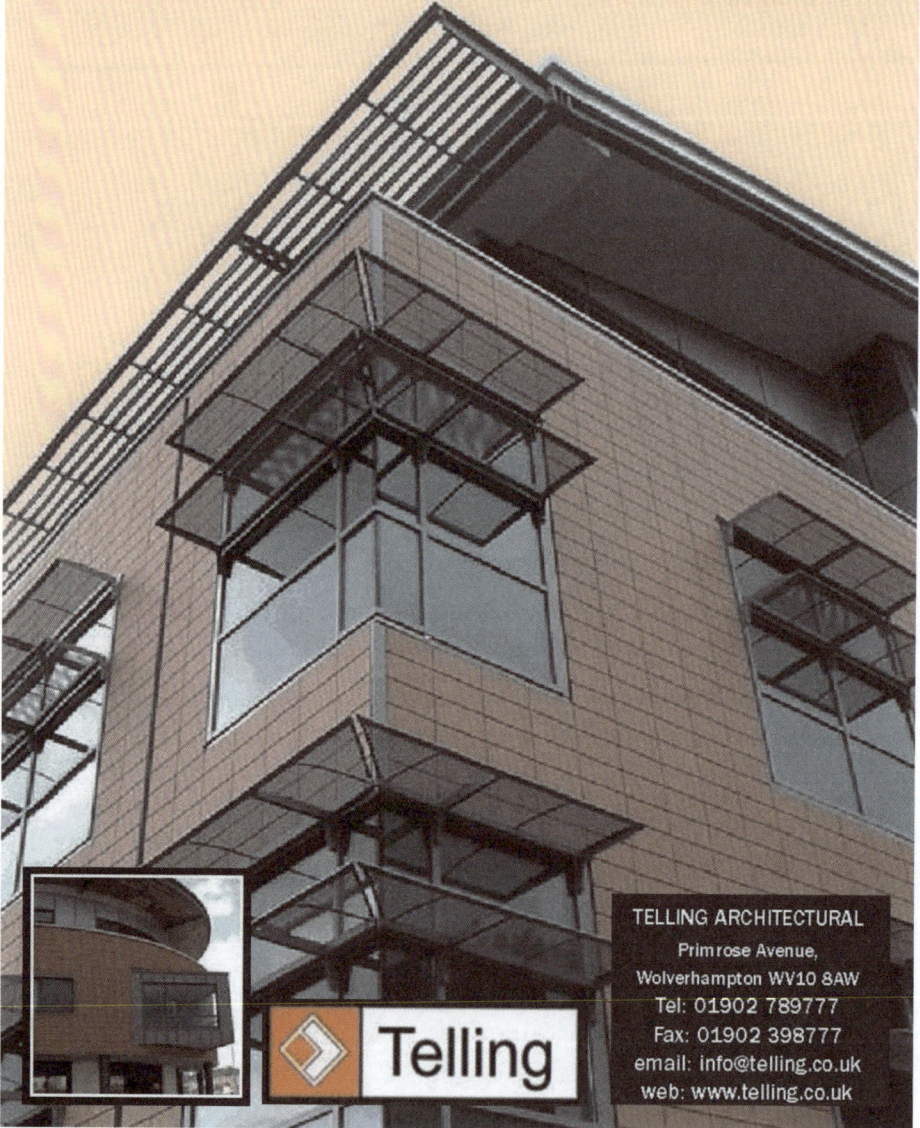

Project: Sovereign Quay, Cardiff-Barratt Homes, Architect: Goddard Manton

Facade   Wall   Ceiling

Facades with looks that are built to last

Sto – 50 years of innovation

Developing and providing products and services for responsible construction work has been the principle of Sto for 50 years now. Take for example StoTherm Classic external wall insulation system.

● Reduces heat loss, achieving energy savings of up to 60%

● Seamless, attractive facades achieved with through-coloured, crack-resistant acrylic render

● The building envelope is protected from driving rain whilst remaining vapour-permeable

● 150 million m² installed worldwide

For more information, call us now on 01256 337603 quoting SOQ/EWI/06 or go to www.sto.co.uk

Sto | Building with conscience.

sto

108

The Eurobrick "Quarry" range provides a stone effect by using reconstructed stone facings in a similar manner to the brick system.

However, the stone effect is formed with "double" course stone slips installed in a manner to replicate a stone facing.

The aggregates used comply with BS 882 for aggregates and include crushed stone, natural sands and gravels.

All pointed joints are coloured to be sympathetic with the coloured stone facings used.

Pointing Brick Slips Pointing of slips, however they are fixed is generally best performed by hand injecting the pointing mortar by gun or bag.

The mortar is specially formulated and coloured for the purpose of completing the slip systems and offers a wide choice of joint type profile.

A specialist gun is used to place the pointing mortar into the joint space, when the initial-set has taken place the surplus mortar is removed and the joint formed into the required profile.

Pointing mortars are portland cement based designed to provide a non-shrink, water-resistant injected mortar. The requirement is for a highly durable finish offering high levels of resistance in exposed situations, including areas subject to wind driven rain and sea spray or high level buildings in harsh climates. Available in a wide variety of colours to compliment the brick colour and texture.

Can be supplied pre-blended and pre-packed in bags of 25kg. by Easipoint.

3.9 Simulated Brick Render
Featured Specialist System BRICKREND
by Kilwaughter Chemical Co Limited

The use of simulated brick finishes is best evaluated when the economics of using brick slips have to be avoided due to budgetary restraints. Realistic results of a true "brick" appearance can be achieved using renders providing the "brick" layout and effect are carefully planned and executed.

Simulated brick render is a cement-based, self-coloured and

Setting Out Courses

Setting Vertical Joints

ready-mixed render requiring only the addition of water.

It is applied preferably in two differing coloured coats, the top coat being cut through to expose the mortar layer, thus creating a brick effect.

All patterns of brick-bonding can be cut, careful setting-out is required to achieve best results avoiding "out-of-module" brick sizes.

Coursings, cut joints and brick sizes must be consistent, for maximum realistic brick effects.

Brick Render is a cement-based, self coloured and ready-mixed render, requiring only the addition of water. It is applied in two different coloured coats, the top coat then being cut through to expose the mortar layer, creating a brick effect.

Brick render should be BBA & ISO 9001 approved and conforms to BS 5262 Type /1.

Traditional Brick Effects

Gable showing Movement Joint

Brick-Rend can be applied over most basecoats.

A sample panel, constructed on site, is recommended to ensure that the specifier approves the render colour and texture. This sample is recommended to be kept on site as a reference for the standard of workmanship and design of brick bonding layout.

Materials are manufactured from natural products, and slight shade variations may occur.

The working day areas of application will be dictated by the skill of the operatives, weather conditions and architectural requirements.

The movement joint nosings can be supplied coloured to "blend-in" with the adjacent colour of the simulated brick.

111

3.10 TRIMS & BEADS

The use of beads, trims and flashings vary on a project by project basis dependent on the specific job specific requirements and to complete the project to a high standard of weather-proofness. It is a part of the design of the project and the trims and beads will be influenced by the substrate, the system make-up and the final aesthetic appearance required.

The design location and installation of any trim, bead or flashing should be sympathetic with the visual elements of the project and to enhance the final appearance. Any visual impairment due to poorly located beads or trims is undesirable.

The purpose of trims, beads and flashings is to ensure that the exposed edges, abutments or cappings of systems and finished surfaces are water-tight and durable. The design of any trim, bead or flashing is to shed water correctly and protect exposed edges and ends to normal wear and tear during its lifetime. Exposed faces of trims, with the exception of stainless steel trims, should be coated with polyester powder coatings which provides weather protection and improves the final appearance with sympathetic colours.

All trims and beads can be supplied with or without an impact resistant pvc nosing. The nosing ensures maximum protection of the bead during application and provides an ideal finishing line for the render or textured finish.

Stainless Steel Products Usually specified in the higher quality project, stainless steel offers long term durability wether the grade is 430 or 304. (Ferric or Austenitic) Difficult to powder-coat, the stainless product is supplied in its natural form which can be "dull" or "shiny". The "dull" effect can look as if it is dirty whereas the "shiny" finish is mirrored and can be very obvious when caught by the sun.

Galvanised Products The design is generally as for stainless steel products, however the galvanising should have an additional coating of polyester powder-coating to assist with greater durability and provide an acceptable finish to exposed faces.

Galvanised profiles are produced from tight coat galvanised steel complying with the requirements of BS2989 1982 with an organic polyester powder coating applied under exacting factory conditions and oven baked to produce a hard finish that will stand up to harsh environmental conditions.

Aluminium Products Most aluminium is produced from bauxite, the main supplies of which are found in Australia, South America, India, the Caribbean and Africa. Aluminium has a higher strength to weight ration than most other metals or materials. Aluminium is attacked by the alkaline qualities of cement so must be protected, usually by the powder-coating process, for long term durability.

Accessories such as cills, flashings and composite panels may be formed from flat sheet material and be fabricated from aluminium alloy 1200 or 3004 or 5251 of appropriate temper all in accordance with the latest edition of BS 1470 - "Aluminium Alloys for sheet products." The standard for powder coatings for application and stoving to aluminium alloy, sheet and preformed sections for external architectural purposes, and for the finish on aluminium alloy extrusions, sheet and preformed sections coated with powder organic coatings is BS 6496.

Plastic Products Trims and beads can be supplied in high impact uPVC, offered in a variety of colours and profiles to suit a wide range of situations and dimensions. There are commercially available several suppliers of these components which between them provide practically all designs and colours that can be achieved.

A particular advantage with PVC is the ability to offer special profiles at economical costs, achieving small dimensions which are impractical with roll-forming or press-breaking metal. Plastic being fairly flexible is to be avoided where large spans are required or where the profile demands stiffness, only formed metal can achieve good stiffness satisfactorily.

3.11 Trim Details

The following trims are the standards for the industry, variations are made to overcome individual contractual and design problems that occur from time to time. The use of particular materials, such as stainless steel or coated galvanised steel, is at the discretion of the Building Surveyor.

Featured trims by WEMECO Limited

BASE TRIM Type A (metal)

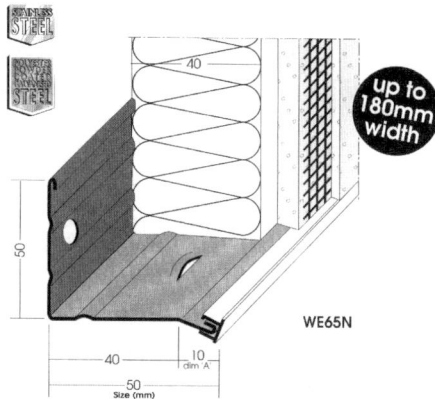

BASE TRIM Type B (Metal)

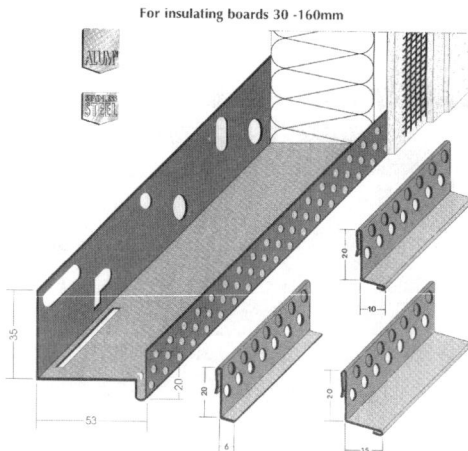

114

End-Stop Trims

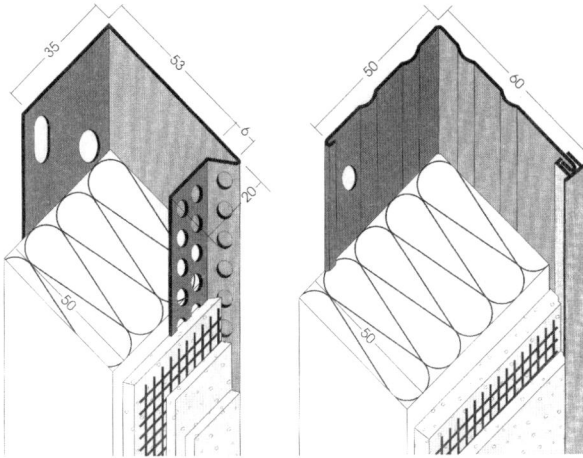

Top-Capping Trim

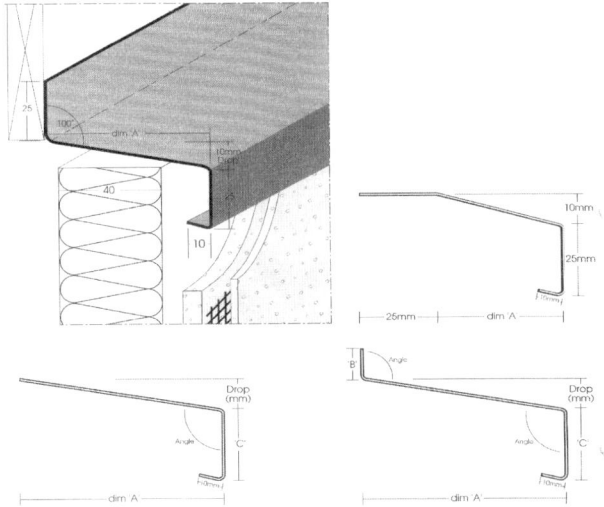

Corner Beads (Expanded Wing)

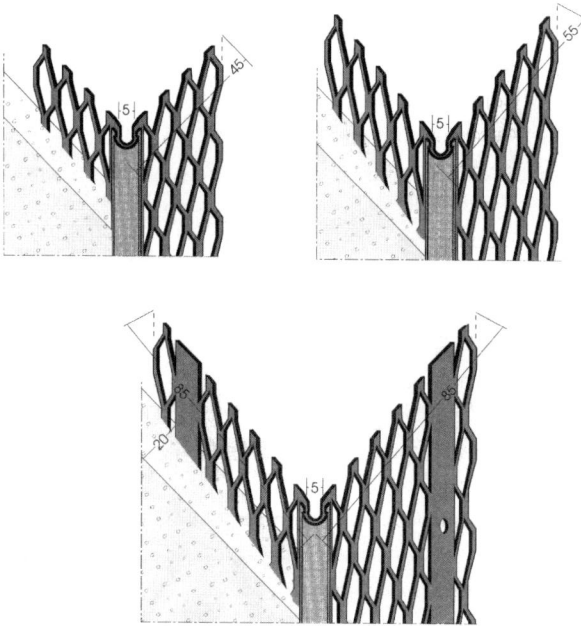

Corner Beads (Pressed Metal)

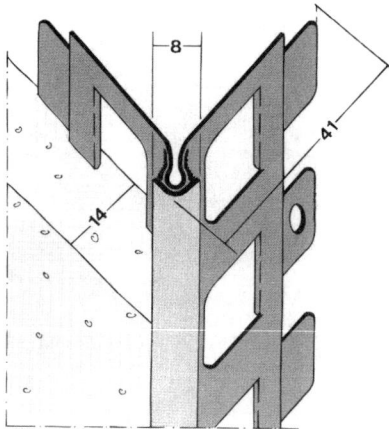

Movement Joint (Render)

Movement Joint Structural

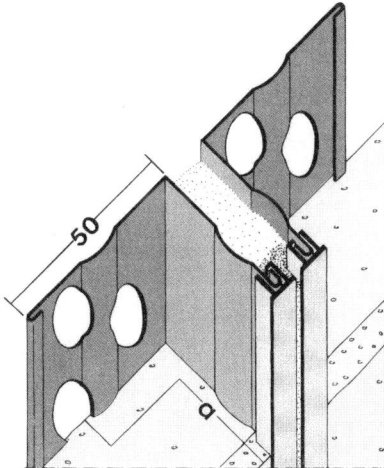

Undercill Trim

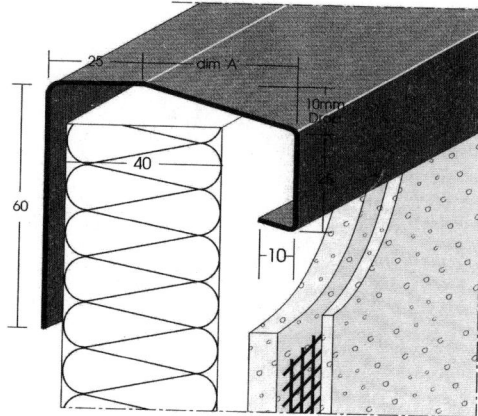

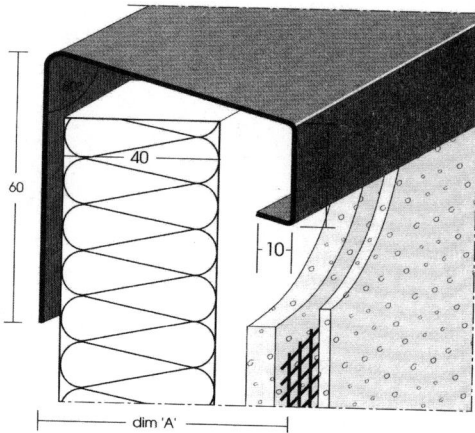

Overcills

Existing Elevation

Fold
Upwards Here

Reveal
Depth

Cut
Here

PLAN OF OVERSILL

Cut
Here

Reveal
Depth

Fold
Here

PLAN OF OVERSILL

Mechanical
Fixing of
Sill into
Reveal

50

Mechanical
Fixing of
Sill into
Wall

50

LENGTH - REVEAL TO REVEAL + 100mm

LENGTH - REVEAL TO REVEAL + 100mm

EXISTING
SILL

Brick or
Wall

New Surface
Finish
System

20

'A'

Drop
(mm)

'C'

'E'
if req'd

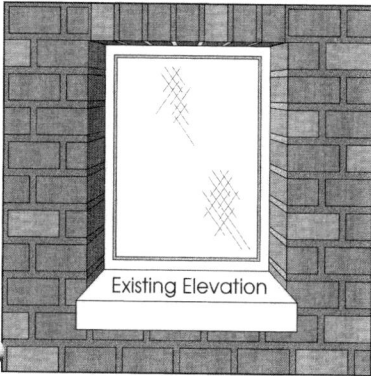

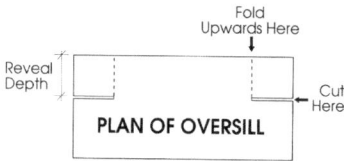

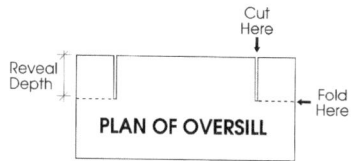

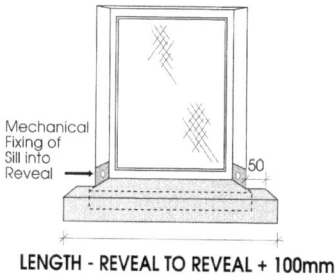

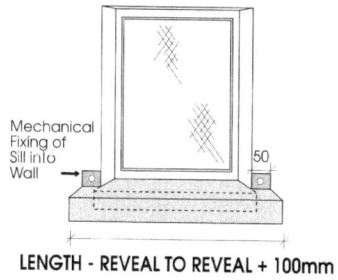

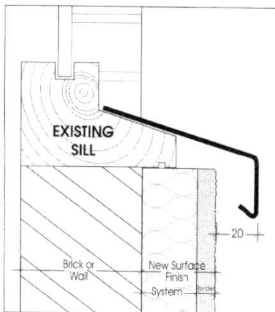

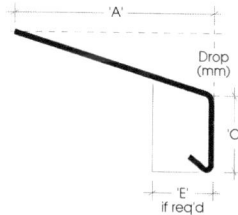

Relevant dimensions (A, C & E)
are taken on site ensuring
dimension 'A' is sufficient to
'oversail' new surface finish by
a minimum of 20mm as shown.
Length is measured reveal to
reveal, + 100mm.
Sills to be fixed prior to
application of new surface
finish, to hide fixing lugs.

119

Custom Fabrication

FEATURE CHANNELS

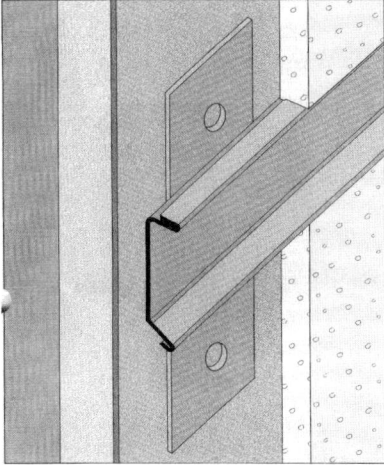

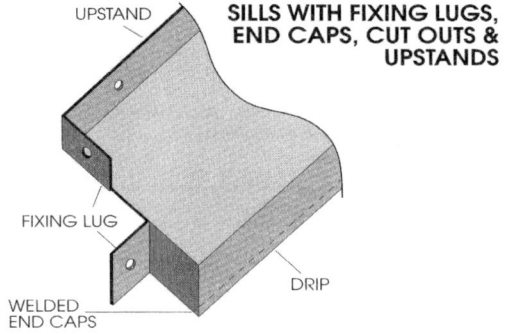

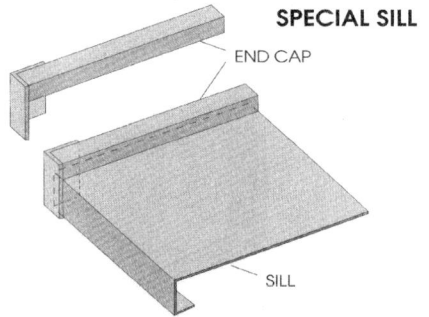

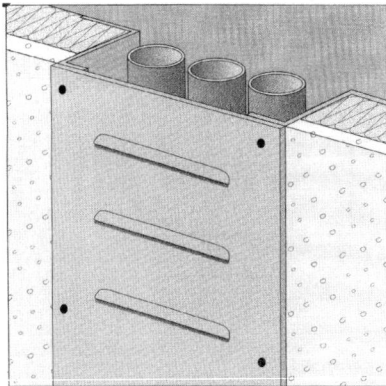

SILLS WITH FIXING LUGS, END CAPS, CUT OUTS & UPSTANDS

UPSTAND

FIXING LUG

WELDED END CAPS

DRIP

SPECIAL SILL

END CAP

SILL

LOUVRED PANELS/ACCESS PANELS

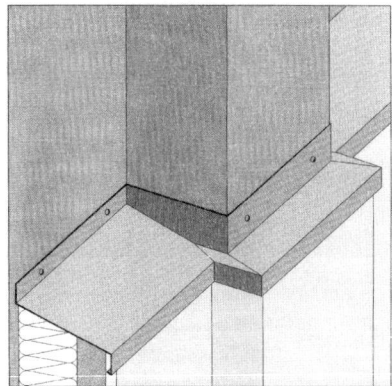

COLUMN SURROUNDS

Curved Trims

Render only

RENDER ONLY
Manufactured from 304 grade stainless steel with optional alternating/one side only fixing lugs.

Example of use:-
To form Architectural curved features - i.e. "permanent formwork" separating renders of differing colours/textures.
To form "edge-stops" to archways where a render is required adjacent to existing arched entrance features.

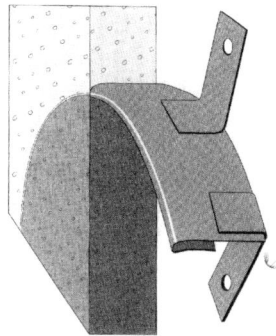

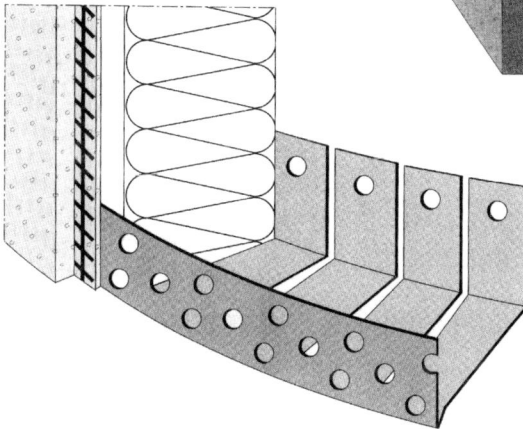

Thermal insulation system Flexi-curve to form curves on plan

Render Stop Beads

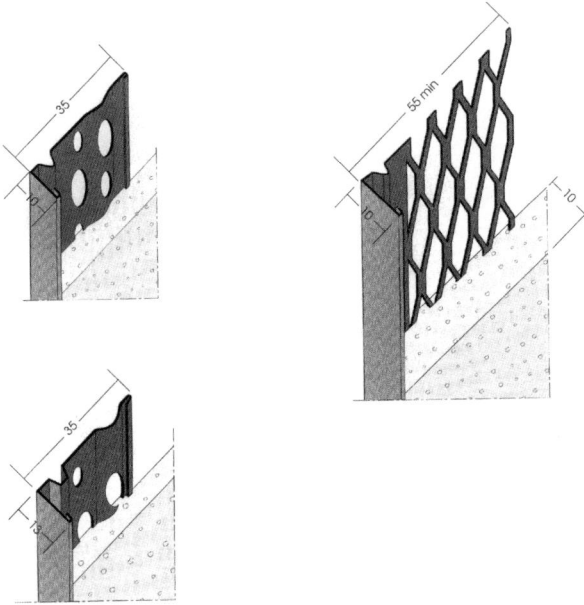

Render Bellcasts

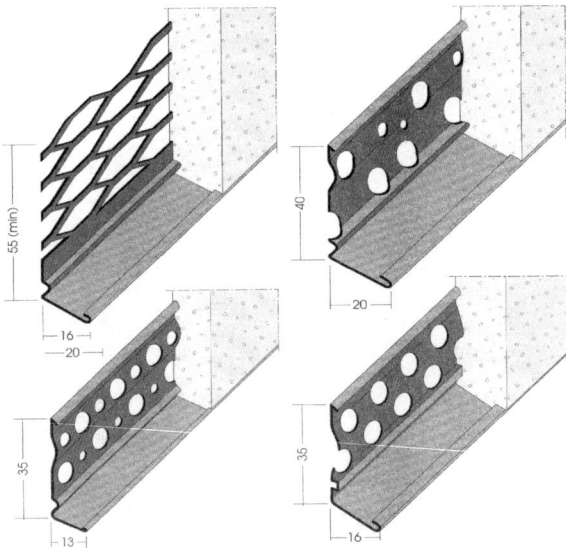

122

Plastic Trims & Beads

Featured Trims by Renderplas Limited

Wide range of sections and sizes for all applications, these beads are available in lengths of 3m, colour white.

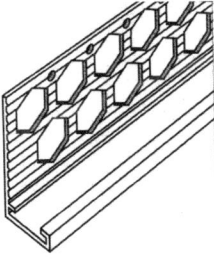

Render Bellcast Beads

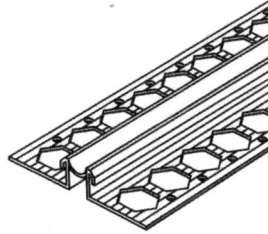

Render Movement Joints

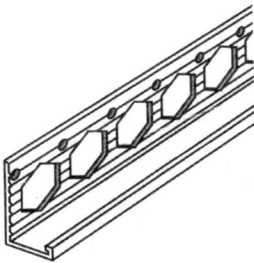

Render Stop Beads

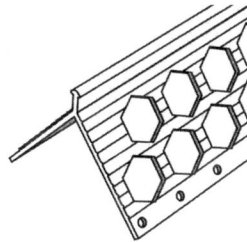

Render Corner Beads

Drip beads

Featured Trims by VWS Technologie AM BAU

Window and door heads together with protrusions from walls such as balconies, require the rainwater impacting onto the surfaces above to discharge itself downwards without percolating horizontally across any soffit. Soffits can be badly stained with contamination if water is allowed to flow in this manner.

To alleviate this problem a drip bead installed into the render at the extreme edge will help prevent this action, thus assisting in water entry prevention and helping to keep the soffits clean.

Below are examples of drip beads suitable for this purpose.

Type 6485 by VWS

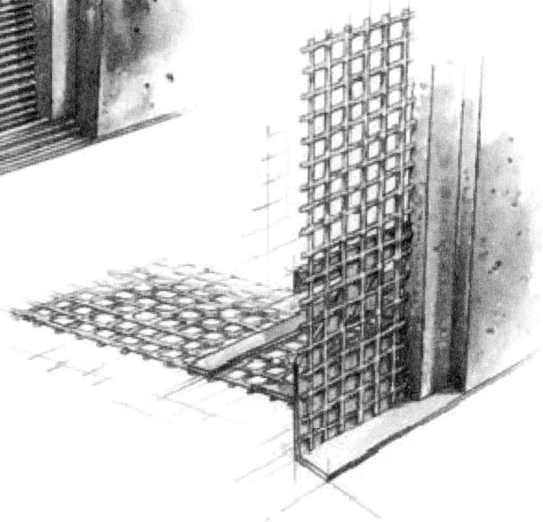

Type 6490 by VWS

Alternative Profiles

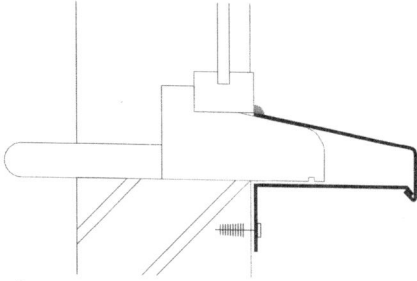

End Features

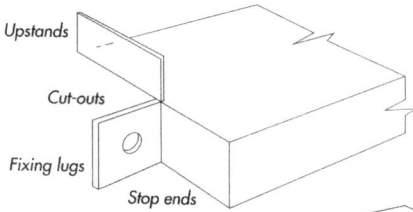

Upstands

Cut-outs

Fixing lugs

Stop ends

ALTERNATIVE PROFILES

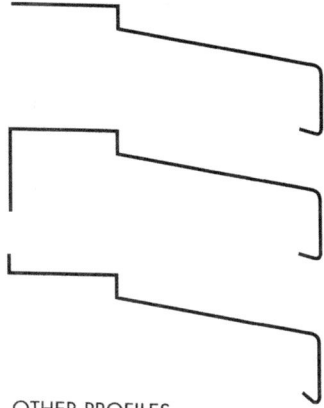

OTHER PROFILES
Available to Clients' Requirements

End Caps

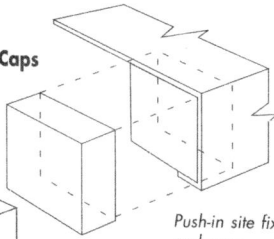

Welded end caps

Push-in site fixed
end caps

Retention Clips

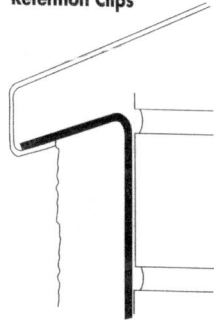

Trims and Flashings

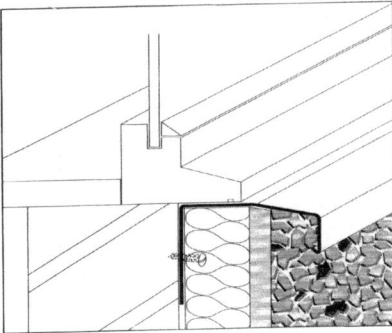

WE 73
(Undercill)

WE 74
(Soffit Flashing)

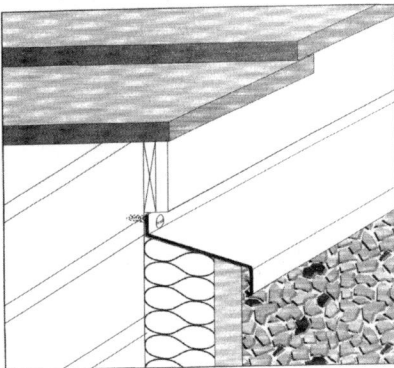

WE 72
(Verge Flashing)

126

Accessories

Aluminium Cills, Flashings & Copings

Generally mill-finished aluminium is capable of being cut, bent and welded in a very wide variety of shapes and profiles to solve the designers problem of protecting the project. The main items constructed are cills, flashings and copings all of which are usually supplied purpose designed and manufactured and with a powder-coated finish to a wide variety of colours.

All aluminium products should be well protected from alkaline contact such as cement as aluminium is easily corroded by such contact.

Pre-painted aluminium can be utilised in a wide variety of profiles but cannot be welded as the heat action of the welding process destroys the coating. Where this is used, joints shall be provided with loose joint plates and sealed with mastic.

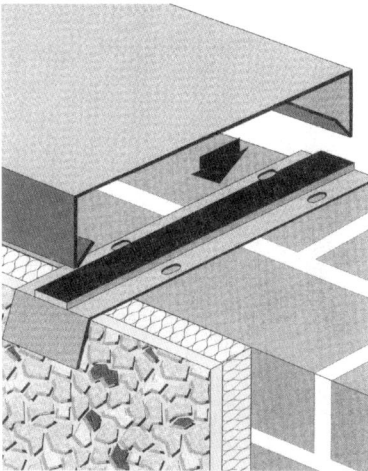

Copings

Aluminium copings provides an economical, weather-proof and easily installed finish to parapet and balcony walls. Copings are fabricated from aluminium sheet or stainless steel in varying thicknesses.

Fixing straps are installed directly to the substrate complete with neoprene seals and joints. All copings can be polyester powder coated with all manufactured joints seam welded for full weather-tightness.

Aluminium copings have been satisfactorily used in many exposed locations, ensure fixings are designed to withstand extreme wind forces, particularly on high-rise buildings.

Allow for thermal movement at joints of approximately 2-3mm.

Fully fabricated joints, corners and abutments can be designed and supplied to suit all locations.

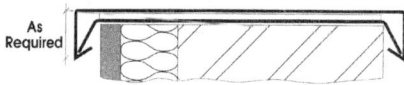

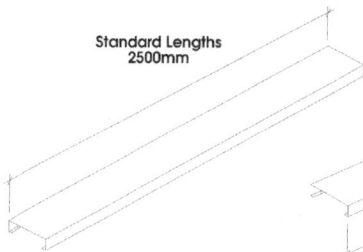

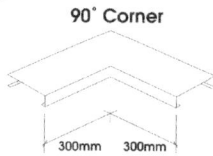

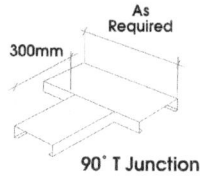

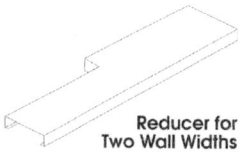

As
Required

Standard Lengths
2500mm

90° Corner

As
Required

300mm

90° T Junction

300mm   300mm

300mm

As
required

Reducer for
Two Wall Widths

Standard Length
with Welded Stopend

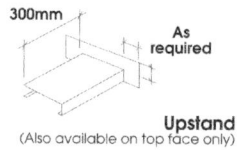

Upstand
(Also available on top face only)

3.12 Mastics and Sealants

External wall insulation systems and renders rely on a variety of mastics or sealants to complete their installation by ensuring the weather-tightness of the project. Generally the mastic or sealant is from one of the following types, namely:-

Gun Grade Sealing Compound An oil based compound intended for the general pointing and sealing of the many joints in traditional forms of construction, rapid and simply applied by hand extrusion gun. Good adhesion to most clean, dry surfaces. It does not sag or slump in vertical joints and possesses good durability and resistance to aging, it can be further improved by painting.

Acrylic Polymer Sealant A one part sealant based upon acrylic polymers. It is gun applied and features excellent adhesion to a wide variety of substrates without the use of a primer. Ideal for jointing between precast concrete cladding panels, perimeter pointing of windows.

Silicone Sealant Generally compliant to BS 5889 1980 One part low modulus sealant which cures on exposure to air to give a seal of very elastic silicone rubber. The sealant is particularly suited to cladding joints subject to substantial moisture or thermal movement and for the perimeter sealing of UPVC and aluminium windows.

Polysulphide Sealant Generally compliant to BS 4254 1983 One part polysulphide polymer based sealant, which is cured by absorption of atmospheric moisture, total curing taking between two and four weeks. Generally has a good adhesion without the use of a primer.

As each sealant has its own particular property and quality it is usually decided by the specifier for a contract as to which sealant will be used, duly considering durability and cost.

Working Characteristics and Performances

Adhesion - Must be good to a wide variety of building substrates including concrete, brickwork, wood, metals and glass, without the use of primers except on excessively porous or friable surfaces.

Flashpoint - 65DegC. Approx

Durability - When used as directed, most good proprietary mastics will provide excellent resistance to weather and give many years of satisfactory service. Regular painting of the compound will improve its durability and extend its life. Note - silicone mastics cannot be painted.

Where buildings are exposed to severe conditions the use of more specialised sealant materials may be necessary and manufacturers should be consulted. Some positions are more prone to failure than others with consequent damage. Historically these are at horizontal positions and south facing elevations, Careful consideration should therefore be given to the selection and application of a particular type and manufacturer of sealants.

Chemical Resistance - Good proprietary mastics should give resistance to normal acidity and alkalinity associated with building materials. However, it is recommended that where there is likely to be an undue concentration of any chemical near the compound, then technical advice should be sort from the supplier or manufacturer.

Movement Accommodation - Movement in tension and compression.

Service Temperature Range - Suggested temperature range to be -25DegC to +35DegC.

Staining - Only proprietary mastics that will not stain are suitable for the use with renders, advice should be sort from manufacturers if in doubt.

Maintenance - The life of most mastics can be extended by regular painting, however, as and when the compound requires maintenance or repair this can only be carried out by raking out the existing sealant and replacing with fresh material.
Note - Silicone mastics cannot usually be over-painted.

Health and Safety Most proprietary mastics are believed to be safe and non-toxic, however, it is wise to avoid unnecessary skin contact and in the event of such contact, to wash thoroughly with soap and water.

If material enters the eye, wash with copious quantities of clean water and obtain medical advice if discomfort persists.

3.13 Render & Component Supply Details

Ready-mix renders Ready-mix pigmented/un-pigmented mortars should comply with BS 4721, supplied usually in bulk to be mixed with Portland cement on site. Mixing to be in accordance with the manufacturers instructions with a precise water-content additive.
Supplied in bulk lots by arrangement with suppliers

Bagged renders The pre-blended proprietary renders are usually supplied pre-bagged in 25kg bags, suitable for site storage in dry conditions. The mix design should comply with BS 5262 and site prepared with only the addition of clean water but in accordance with the manufacturers instructions.

Reinforcements and Accessories Generally all render accessories should of a suitable external quality of either stainless steel, galvanised steel or uPVC. Beads can be supplied with or without coloured nosings.

Beads & Stops supplied in straight lengths 2.5m & 3m lengths

Aluminium powder-coated cills usually purpose made to special order

131

Metal mesh reinforcements can be stainless steel or galvanised steel, supplied in sheets or rolls to a variety of weights.

Supplied in sheet form 1.2m x 2.4m - roll form 1m x 5m

Mineral fibre meshes are usually of alkali-resistant coated glass woven to a square mesh size, polypropylene mesh is also sometimes used.

Supplied in rolls 1m x 50m to a variety of weights & sizes.

Fixings All fixings to secure meshes and beads should be compatible with the material to be fixed and of external quality, fit for its purpose.

Supplied boxed in 100s

Bucket Supplied Coatings Usually of an acrylic or silicon base, these coatings are supplied pre-blended in buckets of factory pigmented and sealed. Application on top of a colour co-ordinated primer is in accordance with the manufacturers instructions.

Supplied in buckets 15kg weight

Masonry Paints Coloured features in smooth render are best achieved with the use of masonry paint, these are usually supplied in tins of to be used as directed by the manufacturers.

Supplied in tins of 5litres

Ancillary Brackets & Cleats All ancillary brackets and cleats should be of material suitable for external use and designed to project through the insulant without distress. The fixing of such brackets should also take account of the bending moment induced by the insulation thickness and fixed direct to the substrate. Suitable sealants should be used to seal all render abutments.

Supplied individually to special order

Aggregate Dash Finishes Natural and manufactured aggregates for dry-dash finishes are supplied pre-blended into recognised colours.

Supplied washed and bagged in 25kg bags. (By the tonne)

Renders Traditional renders, silicone renders and natural hydraulic limes are all supplied pre-blended and pre-bagged in 25kg bags (By the tonne)

PART FOUR - Design & Drawings

4   Numerous factors apply to each individual project or contract and require detailed design considerations and/or calculations in order to produce a viable, properly specified and executed project.

The following headings, not in any particular order, should cover the requirements of most projects.

4.1   Condensation

Condensation manifests itself in two differing manners, surface condensation and interstitial condensation.

Surface condensation is well known as it is visible and effects nearly all of us in the car and in the home, particularly on internal cold glass or hard surfaces.

Interstitial condensation however, is not visible and is formed within structures on the interfaces of differing materials.

Condensation is when the air temperature, which supports the holding of water in the form of water vapour, falls below that level to which it can hold that water . It takes place when warm, moisture-laden air comes in contact with a cold surface.

Humidity is a measure of the amount of water vapour in the air, it can be defined as the ratio of water vapour "mixed with" each sample of air. It is usually expressed as the number of grammes of water vapour in each kilogram of air.

Relative humidity (RH) is a measure of how much water vapour is actually in the air relative to (a percentage of) the air's maximum capacity to hold water vapour. The relative humidity of a sample of air can be expressed as :-

$$RH = \frac{\text{Amount of water vapour in the air}}{\text{Air's capacity to hold water vapour}} \times 100$$

By this definition, RH can change in two ways:-

1.   If the actual amount of water vapour in the air changes
2.   If the maximum capacity of the air to hold water changes

When the temperature changes, it follows that relative humidity depends both on the amount of water vapour in the air and on the temperature of the air.

The Dew-Point Temperature is the temperature at which condensation first starts to form.

If air that was not saturated with moisture started to cool, its capacity to hold water vapour would decrease but the actual amount of water vapour in the air wouldn't change. If the cooling continued, the capacity of the air to hold water vapour would continue decreasing until it could hold no more moisture. At that point, the air would then be saturated.

The temperature at which this occurs is called the "dew-point" temperature.

System Design Requirements - Condensation   One of the primary reasons to externally insulate the walls of buildings is to increase the wall temperature to that above the dew-point temperature thus preventing the formation of condensation.

Other ancillary methods to avoid the formation of condensation revolve around ventilation whereby the quantity of water vapour held internally by the air is kept to a low level, sufficient to help prevent this effect.

The insulation system should permit the passage of water vapour to the external air sufficiently to prevent excessive condensation forming on, within or between its components. The amount of water vapour held in the air depends on temperature, the higher the temperature the greater the quantity of water it will hold.

Internal condensation & mould growth

Internal surface condensation manifests itself in the form of moisture which can lead to green mould growth on the surfaces of internal decorations. This effect is usually at low level in the corner of rooms and around window and door frames. Condensation is most prevalent in solid wall constructed dwellings or when the "U" value is low and when ventilation standards are poor or non-existent.

The above photograph demonstrates the effect of internal reveal condensation forming adjacent to a frame resulting in the formation of mould growth.

Internal Wall Condensation Surface condensation occurs when the surface wall temperature falls below the dew-point temperature. Lack of insulation within the wall structure resulting in a low "U" value will allow the wall to lose heat, so the internal warm air laden with water vapour will condense out on the cooler internal wall surfaces.

Condensation can be one of the worst problems that designers, owners or occupants of buildings experience. Dampness and mould growth caused by surface condensation can not only be very distressing to the occupants causing ill health, but can eventually lead to damage in the building itself.

The thermal insulation and ventilation requirements of national Building Regulations aim to reduce the risk of condensation and mould growth occurring in new buildings. However, designers should take great care to eliminate all possibility of problems caused by condensation, particularly in rehabilitation projects of existing buildings.

Considerations of the additional harmful consequences of condensation include illnesses such as asthma, bronchitis etc..

Thermal insulation, correctly positioned within specific building elements, combined with adequate heating and, where appropriate, the necessary water vapour control via ventilation, should ensure a trouble free design.

Condensation is not a newly discovered phenomenon, but one that has always occurred given the necessary conditions. However, all buildings and in particular houses, are generally more sensitive to condensation now than in previous years. This is due to changes in building design, occupancy patterns, ventilation methods, fuel costs and maybe climate change.

Homes tend to be heated intermittently and moisture producing activities are concentrated into relatively short periods of time.

Ventilation can assist greatly in the control of the quantities of water vapour held in the air. It can be provided in the usual forms through windows which either can be opened or are provided with trickle vents. The action of opening windows relies on the human approach whereas trickle vents are permanently provided and can be more controllable.

The more controllable method of handling ventilation is through the use of heat-exchangers. These provide a method of extracting water-vapour to the outside air with very little loss of heat. The outgoing heated air is used to temper incoming cold air, subsequently heated via a central heating boiler.

Condensation occurs due to the amount of water vapour that air can hold until, at the dew-point it becomes saturated. At any given temperature, air is capable of containing a specific maximum amount of water in invisible vapour form. The warmer the air, the greater the amount of water vapour it can contain. Conversely, the lower the temperature, the smaller the amount. Water vapour in air exerts a pressure, called the vapour pressure. Any differential in vapour pressure causes vapour to diffuse from high to low pressure areas.

Warm air inside a building usually contains more moisture than external air, due to occupants activities, or from the evaporation of residual moisture in new construction. This creates a pressure differential across structural elements. The internal air, being at a higher pressure, tends to diffuse through the structure towards the colder, low pressure, exterior. If moisture laden air comes into contact with a cold surface, it will cool. As it cools, water is subsequently deposited on these surfaces, in the form of condensation.

In addition, warm moist air will always try to diffuse through a building into colder rooms, such as poorly heated bedrooms and stair-wells. This is one reason why surface condensation does not always occur in the room where water vapour is produced.

Interstitial Wall Condensation   Warm moist air will also try to diffuse through building elements to reach colder, lower pressure conditions outside.

If the building materials have low water vapour resistance it is possible for condensation to occur within the building element. In reality, this will be on the first cold surface, at or below dew-point temperature, which is encountered by the moisture vapour on its passage through the structure.

As an example, for double skin masonry walls with cavity wall insulation, the position for condensation to form is on the inner surface of the outer leaf whether or not insulation is included in the cavity.

There is little or no evidence to suggest that interstitial condensation will occur within the core of building materials under normal building and climatic conditions.

4.2   Cold bridges

Masonry walls with or without a cavity, suffer from "Cold-Bridges" when insulation is introduced either externally, internally or within the cavity. These cold-bridges are the result of insufficient attention to the correct detailing required whether by neglect or due to cost restraints when constructed. A "Cold-Bridge" is defined when the continuity of the insulation or the provision of an uninsulated component is broken such that the temperature of the area affected falls below the dew-point causing the formation of condensation. This condensation may be on any surface and is visible or it may be within the structure, called interstitial condensation, which is invisible until damage is apparent.

There are several common areas where condensation may occur, these are notably around windows and doors where the provision of adequate insulation is the most difficult. Condensation can cause damage if it forms on the inner surface of unventilated impervious cladding or if water vapour is restricted from passing through the construction by a vapour resistant layer on the cold side of the insulation. It can also cause damage within the construction if water vapour is unable to permeate through the wall or if moisture laden air reaches the cold surface of the masonry.

In timber framed constructions, if water vapour from inside the building is allowed to pass through the insulation to the cold part of the structure, it may condense. If this occurs for extended periods, the timber structure may be at risk of decay.

The most common problems of "Cold-Bridges" occurs when insulation is installed in refurbishment projects which have severe cost restraints applied, thus the more costly detailing avoiding such problems are generally omitted as an acceptable "risk". These areas are generally as follows:-

Insulation

Floor

Cold-Bridge

G Level

Wall

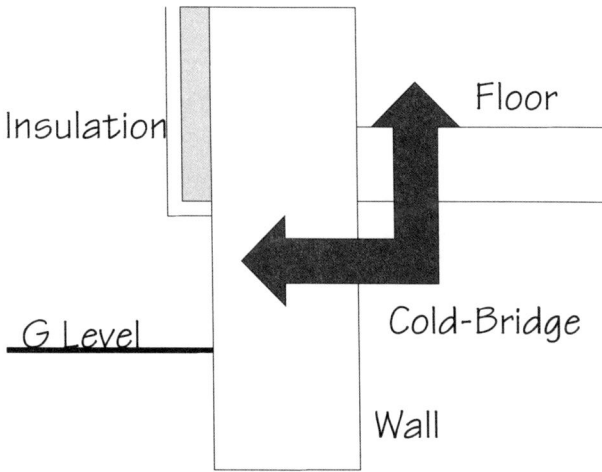

Ground floor junctions to Outside

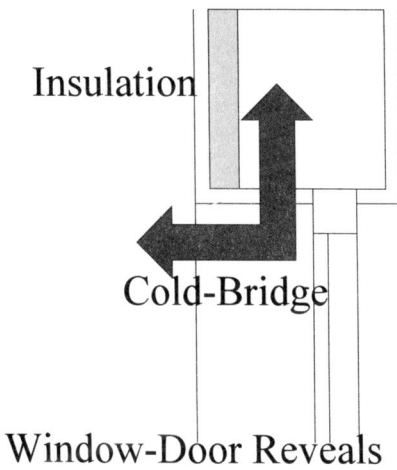

Insulation

Cold-Bridge

Window-Door Reveals

Window & Door Reveals/ Heads

Party wall

Cold-Bridge

Wall

Insulation

End of insulated areas

Roof

Insulation

Cold-Bridge

Wall

Insulation

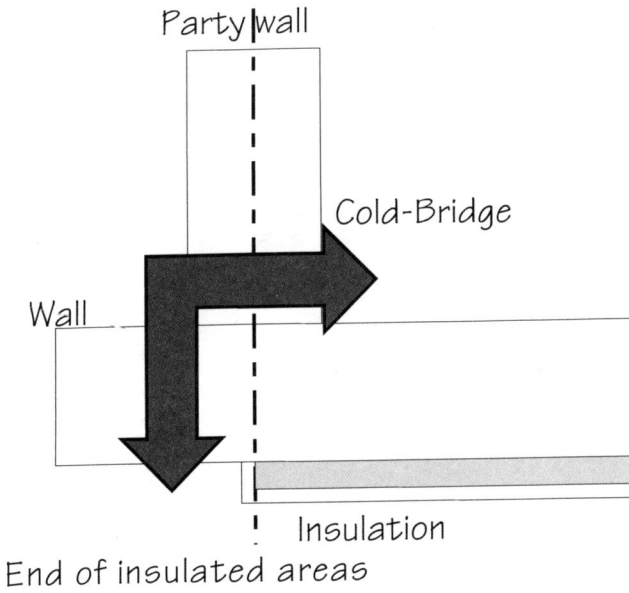

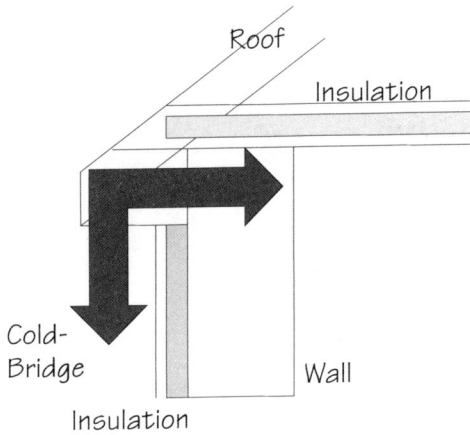

Roof junctions at eaves level

4.3   *Heat loss through floors*

Solid or hollow floors, close to ground level will lose heat to the outside, the quantity of which will depend on the areas of walling of the exposed face. The greater the distance from the exposed wall the less the heat loss.   The greatest problem of heat loss in this respect is from solid concrete floor slab laid directly onto a hard-core base which is in direct contact with the ground support. Values used to determine actual losses are fairly consistent as these do not vary with concrete thicknesses etc..

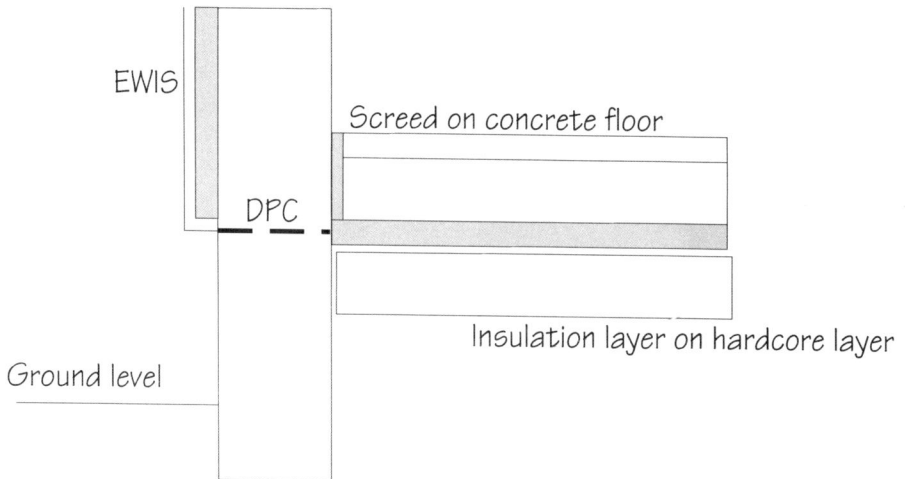

EWIS

Screed on concrete floor

DPC

Insulation layer on hardcore layer

Ground level

INSULATED FLOOR

Condensation can be a problem to the perimeter internal floor surfaces whereby insulation may become desirable.

There are two methods of solving this problem, one is to continue the external insulated cladding down below the damp proof course level, into the ground sufficiently to remove the likelihood of such condensation forming, the other is to insulate the floor itself. This may not be practical in many projects due to the nature of existing construction and cost restraints, the external option may therefore prove most desirable.

When installing insulation below damp proof courses use a closed cell insulant such as extruded polystyrene with the render split with a suitable horizontal bellcast trim. This will assist with keeping the finish of the system clean with the reduction of rainwater splashing.

The addition of some form of drainage such as a "french drain" may also be advisable to ensure excessive water/moisture is restricted from contacting the external wall insulation system.

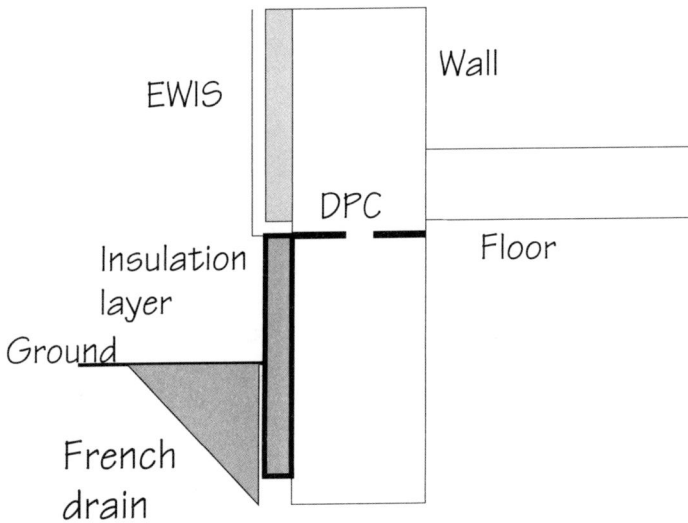

External Drain to System below DPC

Where the insulation detail below damp proof course is specified, the insulation must be constructed sufficiently deep, below ground level, to ensure that the expected condensation risk is eliminated. This is usually a minimum of 300mm below finished outside ground level but can be deeper where conditions demand.

The render used over this section of insulation may be more prone to impact damage as low level impact is more likely from garden tools, toys and bikes etc..

The render at these levels should, therefore, be reinforced to a greater degree than that for conventional walls. Two layers of mineral fibre mesh or the provision of a metal reinforcement such as stainless steel mesh is the usual recommendation. The render should be waterproof and finished smooth ready to receive a paint finish, usually black bituminous waterproof paint. Silicon enriched renders are ideal in these locations.

Considering the first option of externally insulating, the system below the damp proof course level is below the ground level, by a minimum of 300mm. This area of wall will be located in a damp environment so certain precautions must be taken to avoid future problems as follows:-

1.   Insulant must be "closed cell" (Extruded polystyrene) to avoid the uptake of moisture.

2.   Render must be sufficient to withstand damp conditions.

3.   Render must be robust enough to withstand higher impact values and abrasion.

4.   Drainage is recommended.

The second option can usually only be installed in "new-build" situations or where the floor is being or can be removed, perhaps for other reasons, in this case the following criteria should be observed:-

1.   Insulant must be suitable in order to carry the floor load.

2.   Insulant must be protected from moisture ingress or be totally water resistant.

3.   Insulant should provide the sufficient thermal resistence for the floor to achieve the desired overall thermal performance.

Where the floor is a timber suspended floor, the insulation is provided either in the form of a continuous layer of semi-rigid or flexible material laid over, under or between the joists, a suspended concrete floor will best be insulated on top of the concrete construction but below the floor screed.

Insulating Concrete/EPS floor systems are available for "new-build" projects to achieve high insulating thermal performance floors.

In a timber floor, vapour barriers should be avoided or carefully designed to prevent a dangerous situation caused within the floor spaces by water or other liquids collecting either by spillage or leakage from another source.

4.4   *Thermal Design of System*

All external wall insulation systems incorporate some type of insulation, which provides the requisite thermal resistance and reduces the potential for the creation of both interstitial and internal surface condensation. The insulant also provides the surface for the application of the external finishes and provides durability. Selecting the external insulation requires that the existing wall construction be determined first, allowing the designer freedom to offer a variety of design solutions.

Generally, the Building Regulations are to be complied with as the final design requirement, which currently stands at 0.30W/m²DegC., If in doubt consult the local Building Inspector.

Other considerations as to the fire-resistance, thicknesses and costs will determine the ultimate system combination to be offered in open tendering situations.

Consult material manufacturers for values of thermal resistence, vapour permeability and fire performances as necessary.

Standard "U" values are calculated from the resistence of the system's component parts which in turn are based on standard assumptions about moisture contents of materials, rates of heat transfer to surfaces by radiation and convection and airflow rates in ventilated airspaces.

Computer programs are available to assist with these calculations.

See BS 5250 and ISO 13788 for methods of calculation etc..

4.5   *Render Design*

Applying relatively heavy renders and roughcast materials so that they adhere to insulation requires the provision of a reinforcing layer adequately supported to carry the weight of the whole render and finish.   It is possibly the most difficult part of the design operation for not only is some kind of adhesion desirable but also the total encasement of a metal mesh is required.

Installing metal mesh to stand off from the face of the insulation in order to be encased by the basecoat render is desirable and is approached in a number of different ways:

1)   By dishing the mesh at strategic fixing points, say with a hammer so that these dents hold the mesh clear off the board.

2)   By use of special "stand-off" fixings which allow a spacing of the mesh from the face of the insulation.

3)   By inserting spacing pieces or thick washers between the mesh and the board itself at fixing points.

4)   By profiling the insulation surface. (EPS only)

5)   By factory coating the insulation with a polymer modified cement slurry.

Whilst   the   above   expedients   provide   the   necessary embedding of the mesh in the cement, they do not, except in the last alternative, provide adhesion to the insulation board since ordinary cement provides only a relatively poor bond to rigid insulation boards.

To improve the adhesion of the whole system polymers may be added either to the cement slurry or to the basecoat render. These include polyvinyl acetates, methyl cellulose and styrenes: the latter two being truly effective.

Manufacturers instructions should be closely followed when applying these substances, under no circumstances should the breathing capabilities of the systems be impaired. This requires controlling the amount of polymer in the mix as there is some evidence that the vapour transmitting characteristic of polymer cement is inversely proportional to the polymer content (all other factors being equal). Manufacturers of proprietary renders have solved these problems so the use of such renders can be implemented with confidence.

Render Design Options The choice of any particular render will depend on the job-site requirements, Architectural, environmental and location considerations together with costs.

The following table demonstrates a selection of the various choices available.

	Reinforcement	Basecoat	Topcoat/Finish
Heavy-weight Render	Stainless Steel)	10mm/6-8mm	Dry-dash
	Galvanised Steel)		Scratch
			Roughcast
Medium-weight Render	Mineral Fibre)	4-5mm/6-8mm	Dry-dash
			Scratch
			Text Coat
Light-weight Render	Mineral Fibre)	4-5mm	Text Coat

Heavy-weight Render Consisting of two or three coats of cementitious render, mixed in accordance with the requirements of BS5262, these renders are usually installed where specific exposure or resistance to vandalism is demanded.

The reinforcement is metal, either galvanised mild steel or stainless steel fixed through the insulation with either plastic, plastic-metal or all metal fixings direct to the substrate.

Medium-weight Render Consisting of two or three coats of cementitious render, mixed in accordance with the requirements of BS5262, these renders are usually installed where average exposure or low resistance to vandalism is demanded.

The reinforcement is mineral fibre or glass mesh fixed through the insulation with either plastic, plastic-metal or all metal fixings direct to the substrate.

Light-weight Render   Consisting of two or three coats of silicone or acrylic render, supplied pre-mixed in accordance with the manufacturers formulations, these renders are usually installed where light exposure or resistance to vandalism is demanded.

The reinforcement is mineral fibre or glass mesh fixed through the insulation with either plastic, plastic-metal fixings direct to the substrate.

Some systems are adhesive fixed only.

Specialist Renders   Many proprietary renders are available for solving difficult locations with specialized products.

Where these problems are encountered contact the specialist render manufacturers who will assist accordingly.

4.6   System Design Detailing

In pursuant with designing to eliminate or reduce "cold-bridging" most projects will encounter the use of beads and trims to produce a workable installation, the following details will help the reader to appreciate these problems and hopefully help solve particular project requirements.

Window and Door Reveals   Often the most difficult area to avoid "cold-bridges" is the immediate areas around windows and doors. Reveals vary in depth and frame clearances permitting or restricting the use of insulation into these important areas.

Insulated reveal

Window and door reveals can have differing problems dependant on the project and contractual requirements. When refurbishment projects require external insulation the facility of using an insulant within the reveals may or may not be possible.

The size of windows with projecting opening-lights or door jambs will dictate the thickness available for any application. It is quite common practice for insulation thicknesses within window and door reveals to be reduced to allow window and door jambs to be retained allowing window openings and door openings still to function satisfactorily.

Insulated Reveal

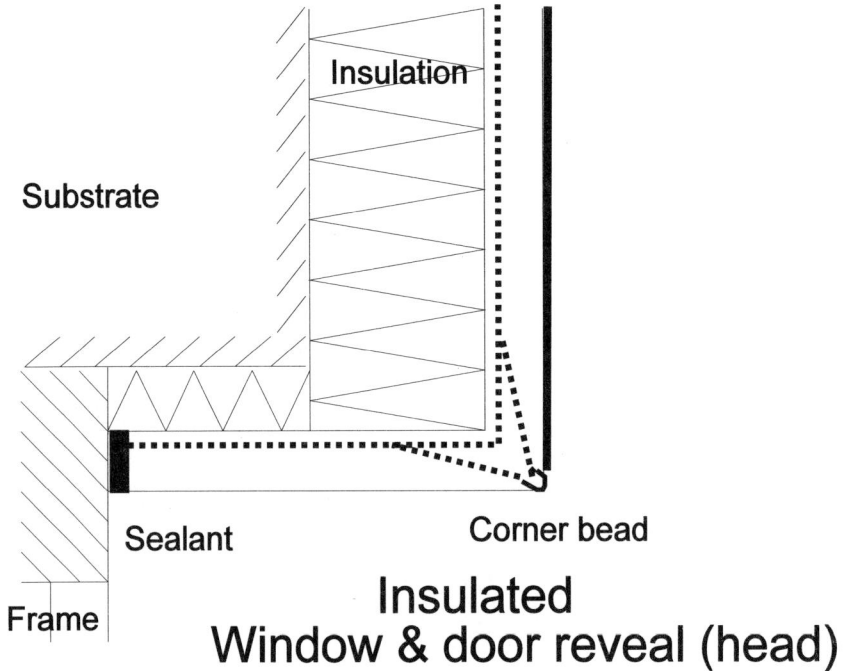

Insulation

Substrate

Sealant   **Corner bead**

Frame

Insulated
Window & door reveal (head)

The insulated reveal can be achieved provided the existing window or door frame can accommodate insulation as shown. This insulation may be reduced from the main system thickness.

Technically, the reduced insulation will not provide full thermal resistence values but may be sufficient to avoid internal condensation forming immediately adjacent to the respective opening.

Un-insulated Reveal

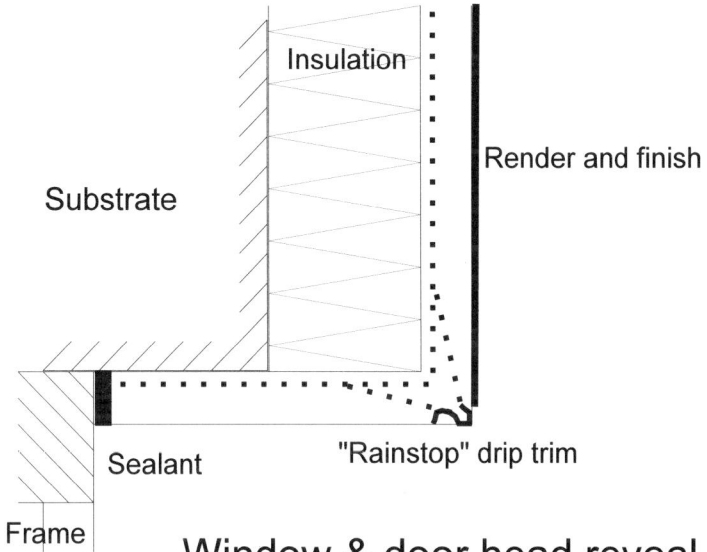

Insulation

Render and finish

Substrate

Sealant

"Rainstop" drip trim

Frame

Window & door head reveal
(un-insulated)

Where a consideration for insulation within a window or door reveal is desirable but not possible, an insulating render may be used in the reveal area.

This will help reduce any "cold-bridge", the source of the problem, without providing the full insulation standard. An economic and practical solution in many situations.

Window Re-Positioned

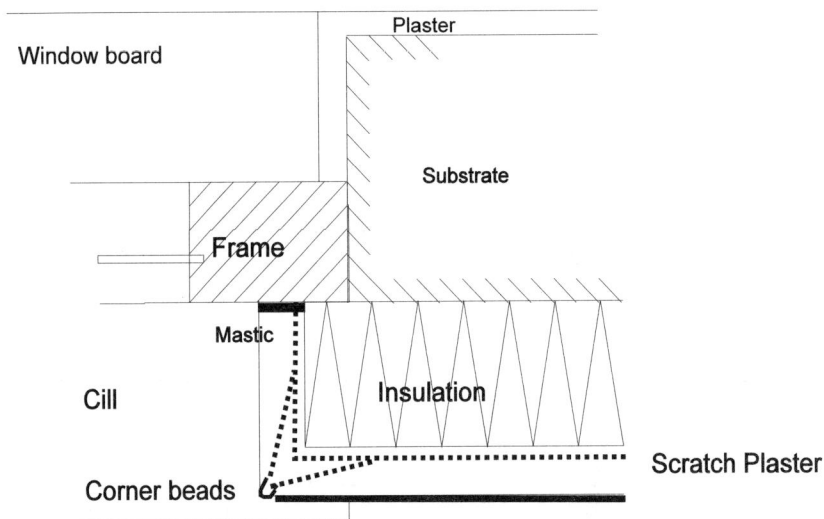

Window board — Plaster — Substrate — Frame — Mastic — Cill — Insulation — Corner beads — Scratch Plaster

Window & door reveal

Re-positioned Frame

The alternative to insulating the reveal or if necessary/possible is to install the window in a new location forward, in order that the frame meets the insulation, reducing the size of the reveal, but completing and forming a continuous insulation layer.

This move will result in more work internally to finish off disrupted walls. The ultimate benefit is that there is no "cold-bridge" and thus eliminating the possibility of internal surface condensation.

This method does have some advantages as the re-use of sound frames can be implemented, thus reducing costs and avoids waste.

Window & Door Frames Replaced

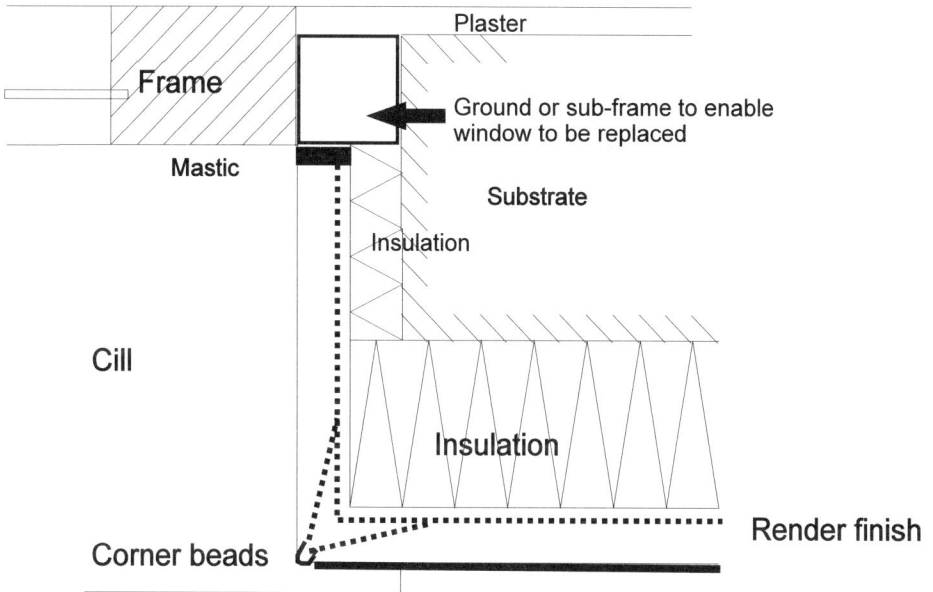

Plaster

Frame

Ground or sub-frame to enable window to be replaced

Mastic

Substrate

Insulation

Cill

Insulation

Corner beads

Render finish

Window & door reveal

Where new windows or door frames are to be installed, the reveal thickness can be determined at the time of construction. The window or door may be provided with a sub-frame or ground in order that the insulation and system may be installed complete before the window or door is fitted.

The size of the timber sub-frame formers can be varied to suite insulation requirements, thus allowing the reveal insulation thickness and size to be selected for optimum performance.

Movement Joints Movement within an external wall insulation system or it's components within it can lead to problems, in the form of render cracks. Consideration should be given to the likely causes and effects of such movement at the initial design stage.

There are four basic causes of movement within insulation systems namely :-

1. Temperature changes
2. Changes in moisture content of the materials
3. The self weight of the materials and external imposed loadings on the system.
4. The effects of both temperature and moisture variations

Temperature changes will occur during the day, from daytime to night-time, from direct sunlight to shade and on a seasonal basis from summer to winter. As a result expansion and contraction will occur in the materials, the extent of which will be influenced by such things as exposure, orientation, thermal capacity, colour and insulation provided by the backing.

As with temperature, changes in moisture levels can occur on a seasonal, daily or even hourly basis and affect differing materials in differing ways. Some absorb water and expand, and some dry-out during curing. Most movement occurs early in the life of the system, however, due allowances must be made to avoid long-term problems or effects.

Imposed loadings generally comprise wind loads, with applied fixtures and fittings being fixed through to the structure. Overcoming the combination of any of the above for design consideration is generally provided for by the inclusion of movement or expansion joints.

Structural movement joints within the supporting structure should not be compromised by the insulation system bridging the joint, the insulation being fixed either side of the joint and a suitable movement or expansion joint extended through the new claddings allowing full movement of the structure without deformation. Due consideration must also be given to movement within the insulation system over wall areas between structural joints.

Renders in particular differ in movement dependent on types and qualities, traditional renders comprising sand/cement mixes generally require movement joints at approximately 5m intervals vertically and horizontally whilst the thinner acrylic renders are relatively free of this requirement. Individual projects should be considered on merit and render performance to decide actual necessity and location.

Movement Joints - Render Only Most cementitious renders require a movement joint to reduce large surface areas to diminish the possibility of cracking and stresses building up on the rendered finishes.

Substrate

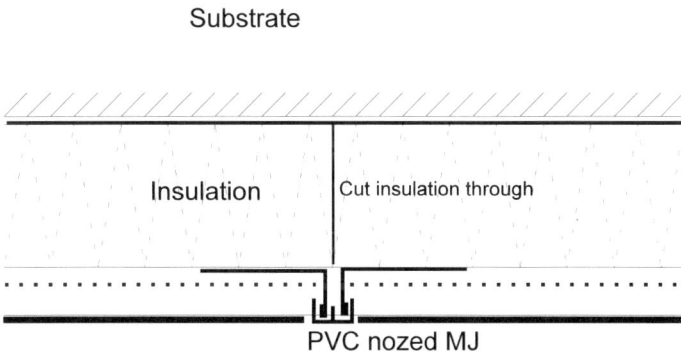

Insulation   Cut insulation through

PVC nozed MJ

Render movement joint

Most system providers recommend that render only movement joints be constructed to avoid dimensions exceeding 5000mm in any direction or areas of render not exceeding 42m².

These joints are formed with a composite bead, adhesive fixed or wired to the reinforcement placed on the surface of the insulant, the render finish ending either side of the bead.

Movement Joints -Structural Large buildings may be constructed with the provision of structural movement joints to allow movement within the structural elements.

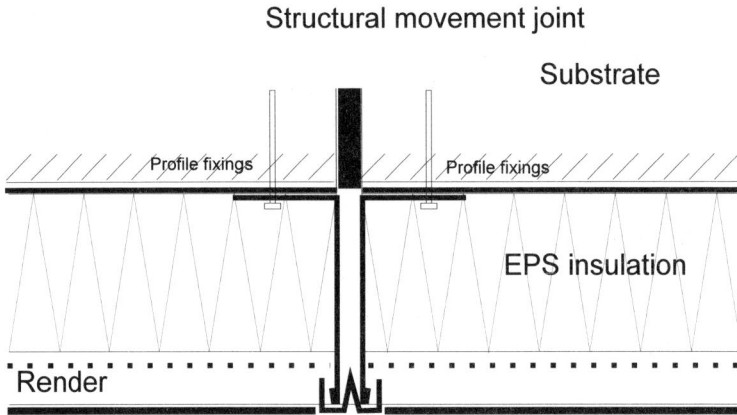

Structural movement joint

PVC nosed movement joint

Structural movement joint

Two types of joint are available :-

1. UPVC nosed
2. Mastic filled joint

Any external wall insulation covering these joints will cause problems if no provision is made to extend the effect of these joints through to the system surface.

Special metal movement joint trims are used to facilitate the structural expansion and contraction through the external insulation system. This new joint should provide equal movement to the original structural joint. The extent of movement will determine the type of joint required.

Abutments System abutments occur on most projects and can involve a variety of detailing to ensure weather-tightness.

The inclusion of a variety of full-depth system stop trims, render only stop beads and a variety of mastics are needed to complete a satisfactory detail.

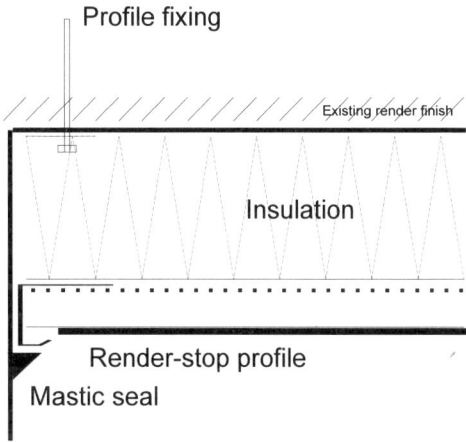

Profile fixing

Existing render finish

Insulation

Render-stop profile

Mastic seal

Abutment detail 1

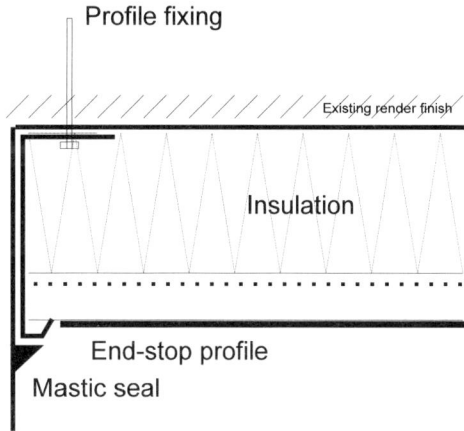

Profile fixing

Existing render finish

Insulation

End-stop profile

Mastic seal

Abutment detail 2

Window Over-Cills   All window cills will require treatment or protection of one sort or another, the two most commonly used solutions are over-cills and under-cills.

Over-cills offer the best solution but at increased costs over under-cills. These are bespoke manufactured powder-coated aluminium cills, made specially for individual window locations, to sizes and designs dictated by the Architect or Project Supervisor.

TYPICAL OVER-CILL

The individual project will determine the detail for attachment

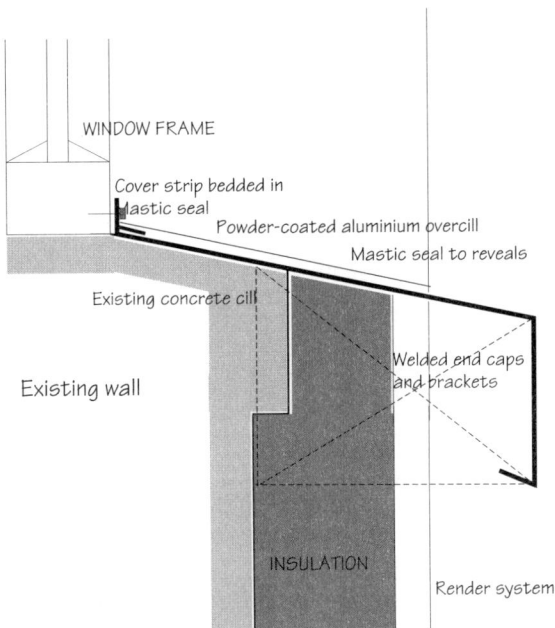

WINDOW FRAME

Cover strip bedded in Mastic seal

Powder-coated aluminium overcill

Mastic seal to reveals

Existing concrete cill

Welded end caps and brackets

Existing wall

INSULATION

Render system

of the cill and the seals needed to ensure weather-proofness. Exposed mastics should be covered where possible, cills should be securely fixed at either end and long cills should be fixed intermittently to avoid any upthrust from extreme weather.

Edge drainage gutters are desirable where possible.

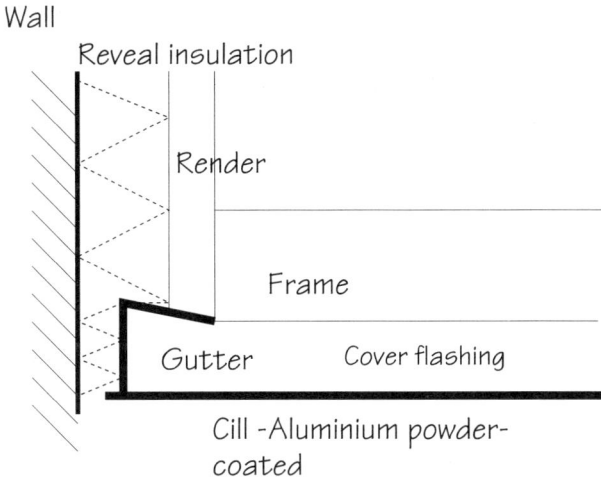

Wall

Reveal insulation

Render

Frame

Gutter   Cover flashing

Cill -Aluminium powder-
coated

SIDE SECTION THROUGH REVEAL

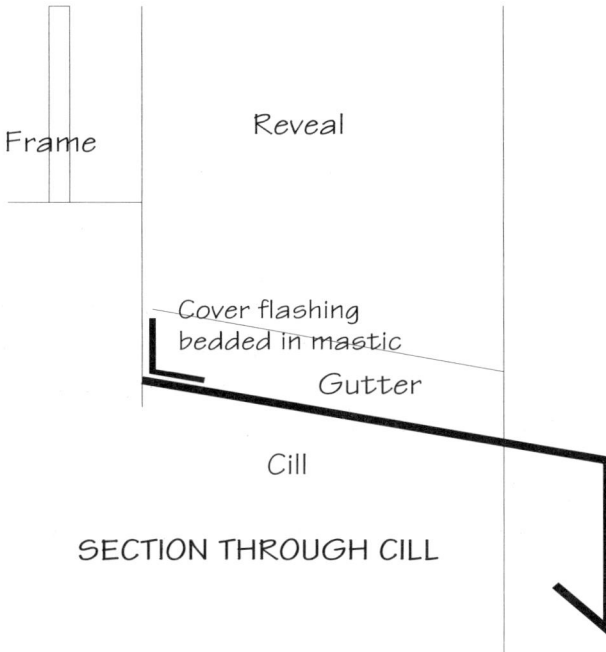

Frame   Reveal

Cover flashing
bedded in mastic
Gutter

Cill

SECTION THROUGH CILL

Window Under-cills   Under-cills allow the provision of a capping to the insulation system and is located below the existing cill. This is the most economical solution but will often not visually appear very acceptable.   The under-cill will usually be supplied in powder-coated aluminium, to a variety of differing colours and fitted with end-caps to seal-off the ends. These can be pre-welded into position or site fixed using silicone mastic.

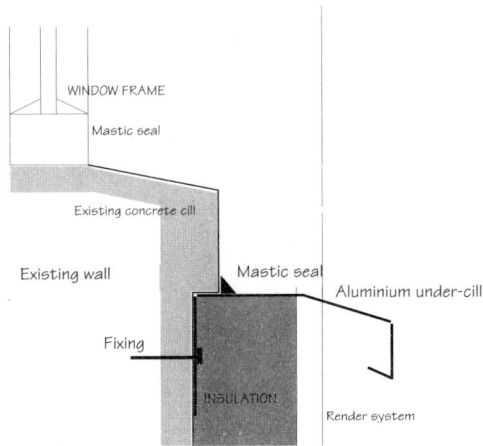

WINDOW FRAME

Mastic seal

Existing concrete cill

Existing wall

Mastic seal

Aluminium under-cill

Fixing

INSULATION

Render system

TYPICAL UNDER-CILL

The mastic seals are most at risk, in the positions as shown but may be the only solution in a competitive situation. The capping trim will be weathered more effectively if bedded into mastic prior to fixing. The edges immediately above the capping trim are vulnerable to water ingress if not adequately sealed.

The old cill left exposed provides a massive "cold-bridge" as no insulation is provided, this should be resisted if possible.

Base Details

Base Detail 1 at DPC Level
(Using combined bellcast)

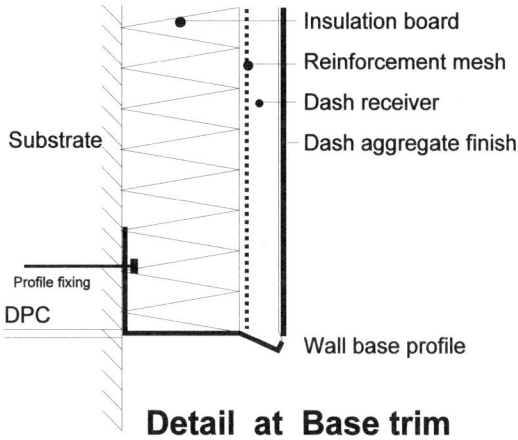

Substrate

Insulation board

Reinforcement mesh

Dash receiver

Dash aggregate finish

Profile fixing

DPC

Wall base profile

Detail at Base trim

Base Detail 2 at DPC Level
Flush base detail

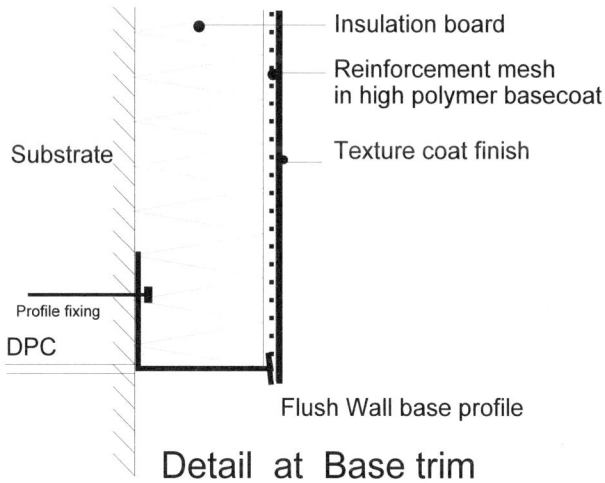

Substrate

Insulation board

Reinforcement mesh
in high polymer basecoat

Texture coat finish

Profile fixing

DPC

Flush Wall base profile

Detail at Base trim

Insulated Base Detail
Either base detail can be used

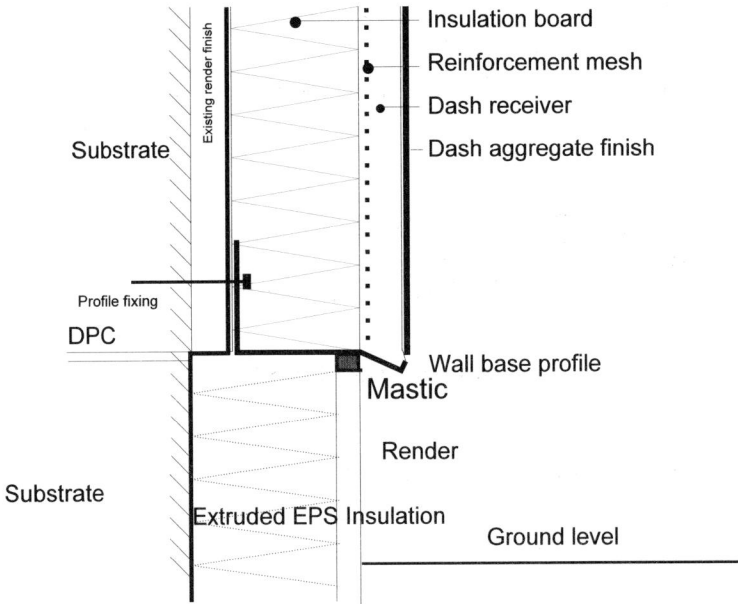

Labels in diagram:
- Insulation board
- Reinforcement mesh
- Dash receiver
- Dash aggregate finish
- Existing render finish
- Substrate
- Profile fixing
- DPC
- Wall base profile
- Mastic
- Render
- Substrate
- Extruded EPS Insulation
- Ground level

Detail at Base trim

The insulated base as above requires that the base trim over-hangs the lower insulation to discharge rainwater away from this lower section. To achieve this when there is no projection render as shown, use thinner insulation even though the insulation values may be reduced slightly.

All base details require that adequate preparation of the wall, prior to installing the trim, is carried out. Any fluctuation or deviation of wall surfaces produce deflections in trims that are unsightly and visually unacceptable.

Eaves Details   Eaves and verges differ widely from house to house and where there is no sufficient overhang, capping trims are required.   These are usually supplied in powder-coated aluminium, to a variety of colours and shapes to adequately cap and seal the external wall insulation system. Mastics are also required to additionally seal the trim to the supporting substrate.

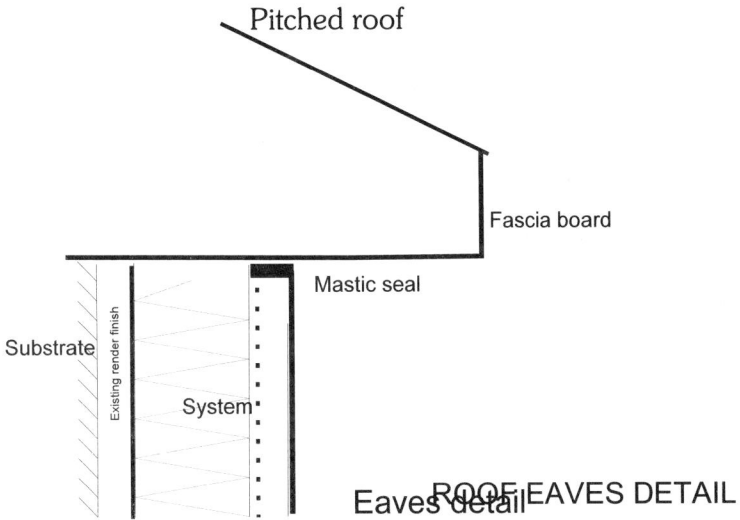

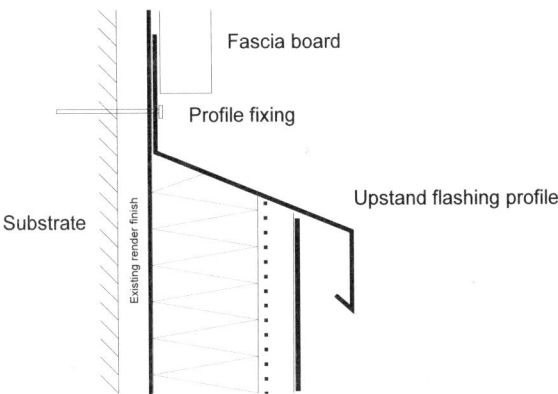

Pitched roof

Fascia board

Mastic seal

Existing render finish

Substrate

System

Eaves detail

ROOF EAVES DETAIL

Fascia board

Profile fixing

Upstand flashing profile

Existing render finish

Substrate

FLAT ROOF EAVES DETAIL

Verge Details   Cladding gables may require the roof tiles to be extended to cover and protect the top of the external wall insulation system. Where this is not possible, for any reason, then the system must be protected with a capping trim.

This capping trim will be manufactured in either stainless steel or powder-coated aluminium.

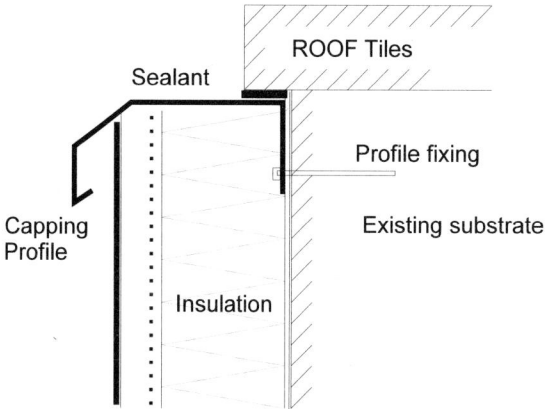

ROOF Tiles

Sealant

Profile fixing

Capping Profile

Existing substrate

Insulation

Verge trim detail 1

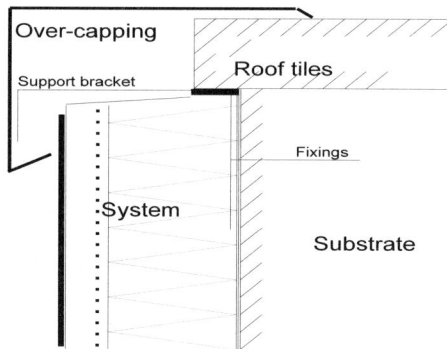

Over-capping

Support bracket

Roof tiles

Fixings

System

Substrate

Verge trim detail 2

Drips & Overhangs Window and door heads traditionally have a drip formed in rendered finishes to help prevent the passage of storm water over the soffit surface. The construction of a drip in sand/cement render was usually carried out with the casting of a groove using a quadrant or bead, this being removed after the render has cured.

Over the years, due to rising costs, this method has been largely ignored with the result that soffits become black with dirt and grime. Cills also require a good overhang to throw water away from the rendered surface in order that this area of wall remains clean and evenly weathered.

To enable a contractor to economically install a drip at window and door heads, proprietary beads can be used to form such a drip or groove, cills or overhangs to trims should be designed to provide a suitable projection, normally approximately 30-35mm. Additional considerations may be necessary for the more severely exposed situations requiring detailed special trims or flashings.

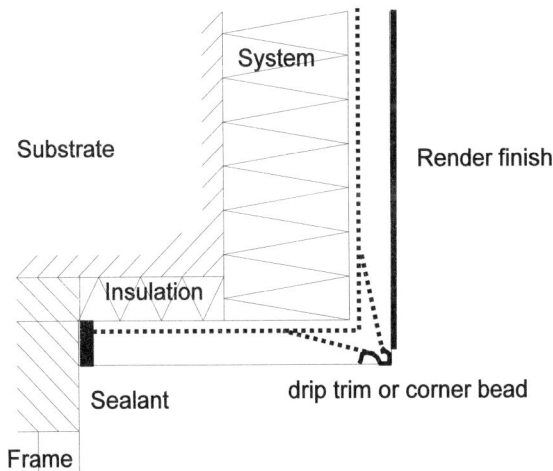

System

Substrate

Render finish

Insulation

Sealant

drip trim or corner bead

Frame

Window & door head reveal

Proprietary drip trims are available

Mid-Point Trims   Where finishes change at floor levels or at "mid-points" then trims are recommended to assist in the definition of demarcation lines. This also assists with the method of application of the differing finishes and in the weather-proofness of the joint.

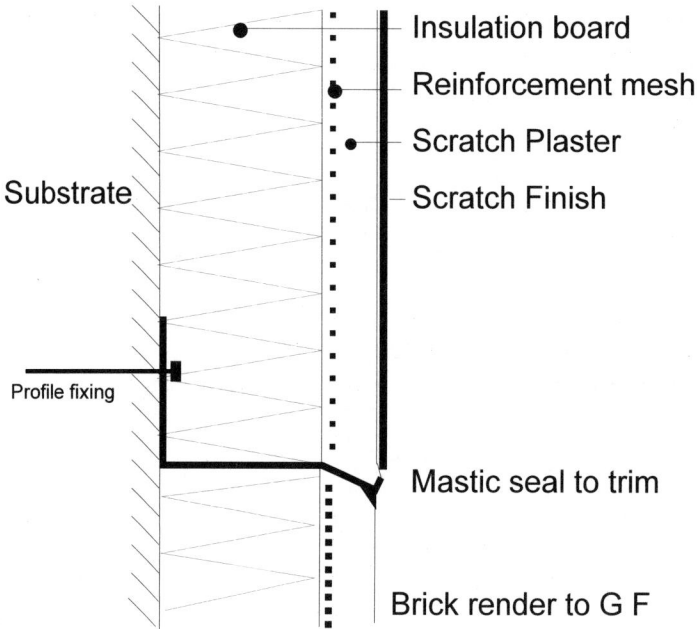

Substrate

Insulation board

Reinforcement mesh

Scratch Plaster

Scratch Finish

Profile fixing

Mastic seal to trim

Brick render to G F

Detail at MID-POINT

This detail is usually installed at a height whereby it is easily aligned by eye, i.e. seen at a level which is easily visually assessed for its horizontal alignment. In order to alleviate any adverse effects of this alignment, it is essential to prepare the sub-strate to carry the trim in an horizontal line which is straight and true, without any deformations.

The render should be applied evenly along the length of the trim ensuring an even exposure of the nosing.

Replacement Window Cills Window cills are an important feature of any project to enhance the project, be waterproof and ensure durability for the surrounding systems by preventing localized deterioration by weather conditions.

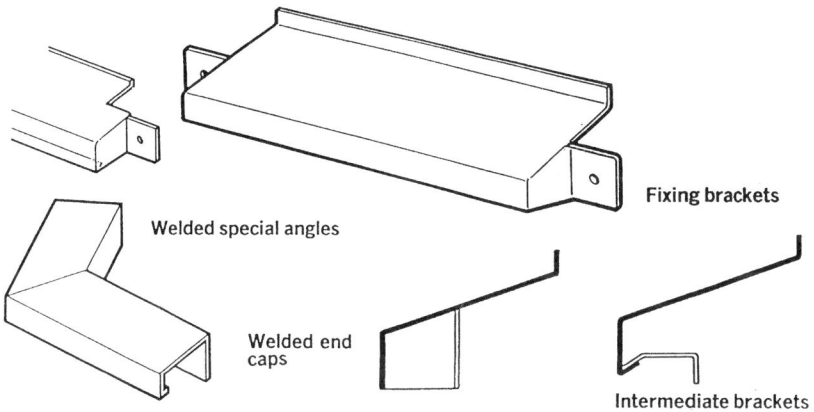

Welded special angles

Fixing brackets

Welded end caps

Intermediate brackets

The preferred manufacture of cills is by construction using 2-3mm thick aluminium powder-coated finish. The construction is all welded with mastic seals at all points of contact with the surrounding finishes. Other details to consider are :-

1.   Edge reveal gutters

2.   Joints in long cills

3.   Colours to be sympathetic with adjoining finishes

4.   Curves

Usually all cills of this type are purpose-made for each project by specialist supplier.

Coping Details Copings are an important weathering feature where parapets are present on projects, these copings seal and protect the system at high levels and are coloured to suite adjoining finishes together with architectural designs and features. Constructed usually in aluminium with a powder-coated finish, welded corners, "T" junctions, curves and size changes can all be accommodated.

The design and construction of the coping should take into account the exposure to wind, rain and upthrusts to ensure long-term weathering performance.

The installation of fixings, sealants at joints providing good

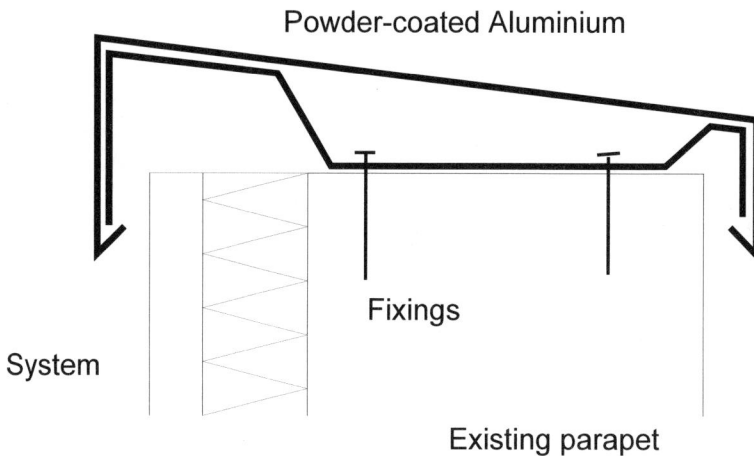

Powder-coated Aluminium

Fixings

System

Existing parapet

Coping Detail

waterproofing and good protective finishes are required for a maintenance free coping project. Details of size transitions, "T" junctions, corners and ends should be planned, designed and agreed before manufacture.

Window Pods

Where designers require to enhance the window as a particular feature, this is possible by using the "window pod".

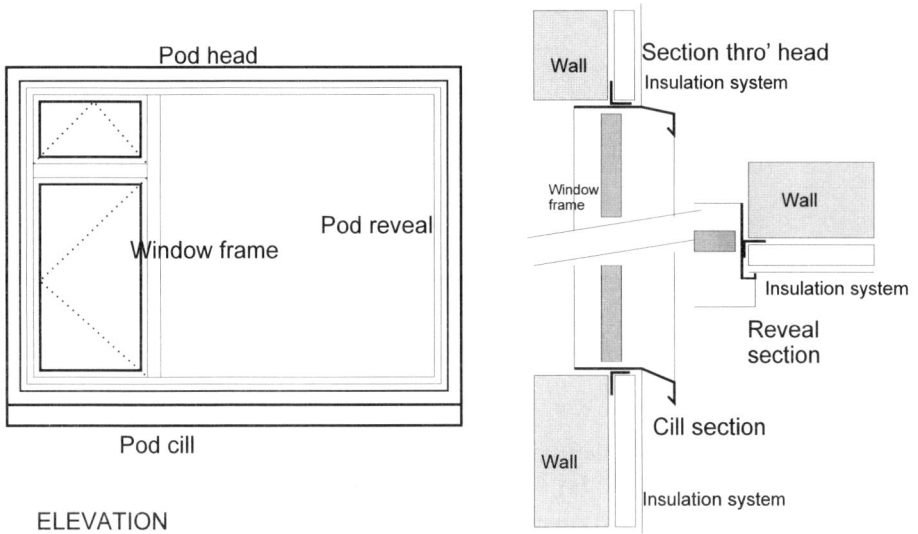

ELEVATION

The "window pod" is a fully fabricated window surround that includes head, reveal and cill. The pod is fabricated in aluminium, usually 3mm thick and is powder-coated to a variety of colours. Almost any reasonable size can be manufactured to specific project details.

The pod is secured fully to the main wall before the insulation system is installed. The window frame is secured within the pod leaving the requisite clearances to be fully sealed with mastic.

The external wall insulation system is installed up to the projecting aluminium pod and fully sealed.

Junction of New & Old Systems   Where buildings are already insulated with an external wall insulation system, which is required to be altered or extended, the alteration or extension must be compatible with the original system used. This is achieved with the aid of special trims, as detailed below.

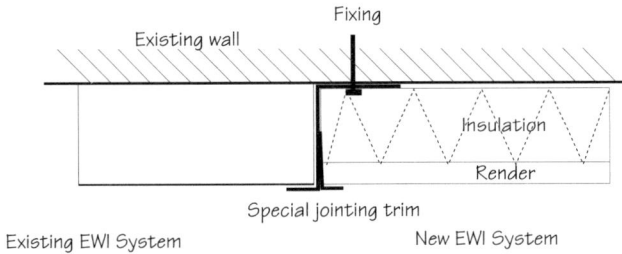

JUNCTION BETWEEN NEW AND OLD SYSTEMS

The existing system is cut with a true and vertical line to allow the new trim to be fixed tightly and sealed if necessary with mastic. The new or extended system is installed from this trim in the usual manner.

Alternatively if a trim is not required then the new and old systems are butt-jointed with a mastic seal to ensure a weather-proof junction.

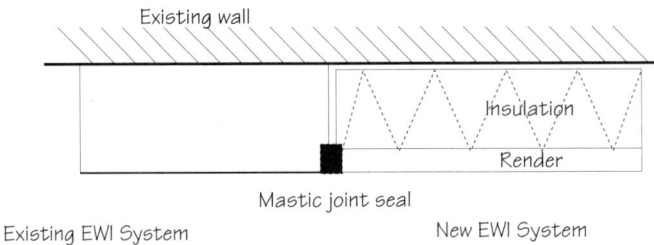

JUNCTION BETWEEN NEW AND OLD SYSTEMS

Service Ducts   Pipes and cables are present on the external face of some properties and may be costly to remove or reposition. An economical solution is to provide an accessible cover, removable for maintenance within the depth of the system. Protruding pipes or ducts may require extended covers, all will need to be specially designed and manufactured.

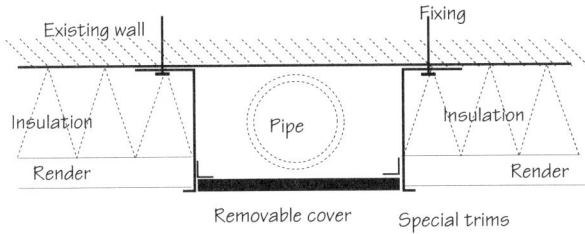

DUCT WITH REMOVABLE COVER

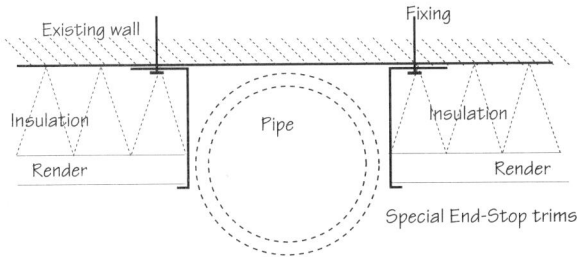

EXPOSED PIPE WITHOUT COVER

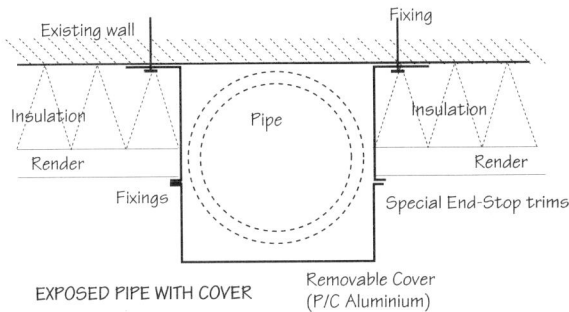

EXPOSED PIPE WITH COVER

4.7   Fire Breaks

Fire-breaks are often required to be designed into a project, particularly high-rise projects where compartment walls and floors are already constructed. It is usual for the compartmentalisation of a building to indicate the location of fire-stops or breaks in order that the function of a compartment wall or floor is transferred into the external wall insulation system.

The construction of a fire-break is usually the insertion of a fire-proof material band, the full thickness of the insulation material for a depth of approximately 150mm or a minimum of the thickness of the system. The conventional method of construction is to use a metal trim to break the system and install the fire-proof material on top of the metal trim. The trim is bedded onto a sand/cement mortar to seal the back-plate of the trim, so preventing the passage of fire behind the trim.

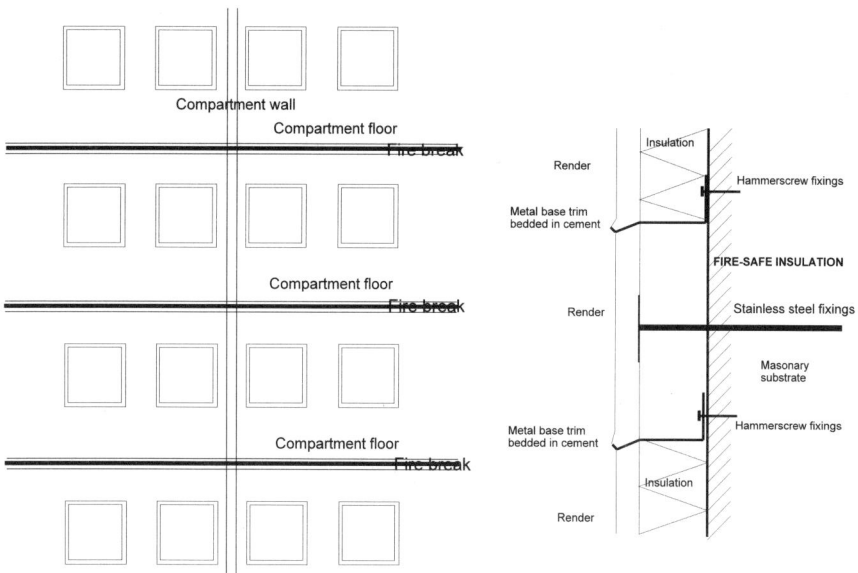

Layout of Fire Breaks

Fire Break Detail

Any difference in insulation material to construct the "fire-break" should not constitute a "cold-bridge"

4.8 Systems with Drainage or Ventilated Cavities

Timber-Framed Systems External Wall Insulation Systems applied onto prefabricated structures can be designed with a drainage or ventilation cavity as a method of installing a second line of defence from water ingress especially in severely exposed locations.

The drainage cavity is a space approximately 15mm wide constructed behind the render or finishing layer and in front of the sheathing of any prefabricated structure or existing wall requiring protection.

As the drainage cavity is ventilated through the peripheral trims externally to the insulation layer, there is little or no loss of insulation value when considering the overall performance of this type of construction.

The cavity construction should include all of the following design considerations:-

1   All cavity internal surfaces should be Class O or "fire-safe"

2   The spacers should be incombustible and evenly spaced

3   The drainage should be direct to the external surface without deflection

4   Drainage through trims shall be provided with trims with suitable perforations

5   Fire-Stops should be of metal construction and coated with intumescent paint

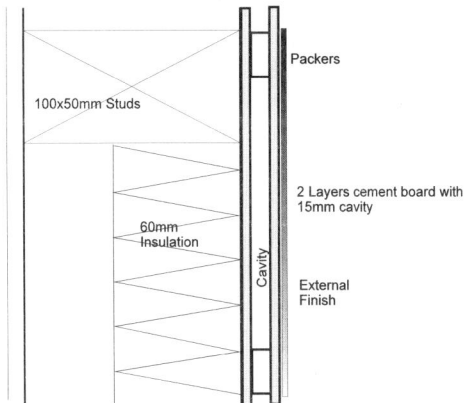

Insulated Drained System Severe exposed sites or extreme heights may require a project to consider an integral drainage system constructed onto a building. This can be achieved in the following manner as the drawing illustrates.

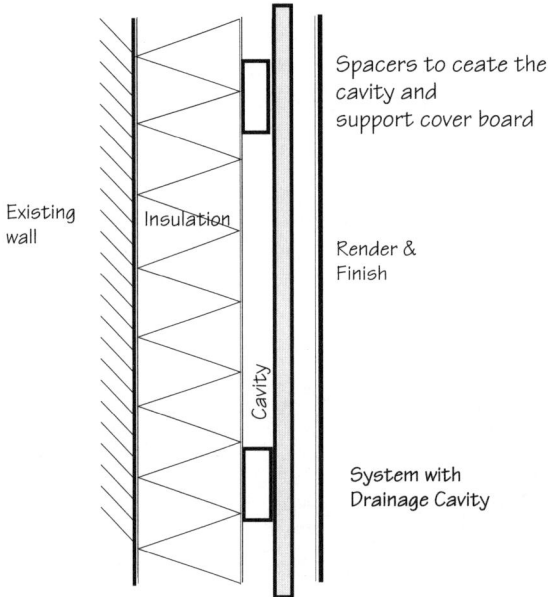

Existing wall

Insulation

Cavity

Spacers to ceate the cavity and support cover board

Render & Finish

System with Drainage Cavity

Notes for cavity drainage systems

1. Spacers to construct the cavity are to allow water drainage vertically

2. Base of system trims to be perforated to allow water to drain through.

3. Drainage required at window & door heads.

4. Fire stops may be required on multi-storey projects.

5. Mechanical fixings penetrate the cavity into the substrate to secure the system.

4.9 Architectural Features

Many projects due to their location or status will require the provision of various architectural features such as those demonstrated below. These features can be created as shown and will assist the project manager to provide the planners with the sustainable detailing that is required.

Quoins Quoins, or corner projecting features are a traditional embellishment to systems particularly where projects are undertaken in conservation areas or areas of traditional building styles. The building owner may require these features to be replicated on the project to be insulated. Quoins can be constructed in render or brick slips, the sizes can be varied to suit adjacent quoins but brick slip quoins are tied to brick sizes and coursing. Stone quoins can be replicated with "stone-effect" render.

The method of construction with rendered quoins is to add additional layers to the conventional finish and shape the arises to match existing quoins or to any specific design. Extra thick brick slips can be installed to conventional coursing and brick sizes, with the adjoining finish abutting the featured quoin.

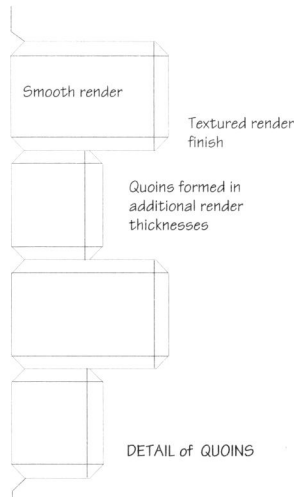

Smooth render

Textured render finish

Quoins formed in additional render thicknesses

DETAIL of QUOINS

Typical quoin feature

After the basecoat is applied, extra render coats are applied to create the desired effect as shown, the main wall finish being applied after completion of the quoins.

Alternatively, quoins can be applied to corners in a pre-fabricated form supplied manufactured in concrete, to be finished and decorated on site.

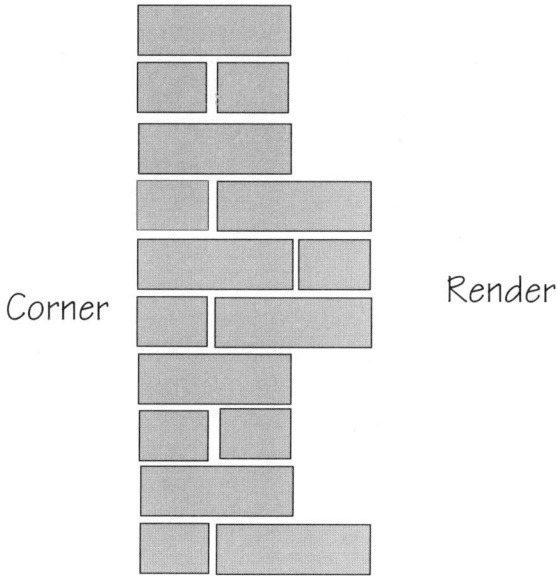

Corner   Render

Brick slips are also used to provide the "quoin" feature to corners of rendered buildings. Extra thick brick slips can also be installed to conventional coursing and brick sizes, with the adjoining finish abutting the featured quoin.

Brick patterns and styles can, of course, be varied to suit design requirements.

Bandings   Bandings can be projecting or recessed, dependent on the Architect/Building Owner requirements but should be carefully considered as any band feature, projecting or recessed can create a weathering problem.

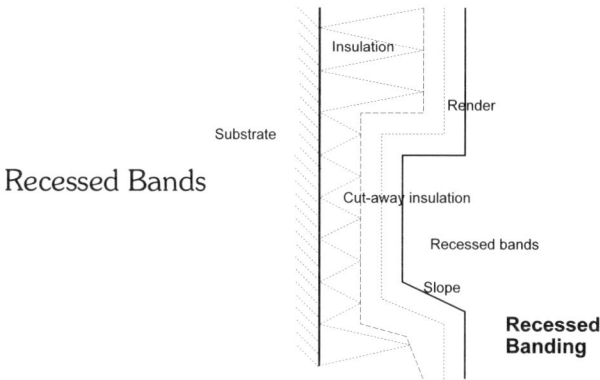

Recessed Bands

Insulation
Render
Substrate
Cut-away insulation
Recessed bands
Slope
Recessed Banding

The recesses are cut into the render after the initial-set has taken place. In this way the bandings can be constructed with good sharp arrises. Beads may be used if desired.

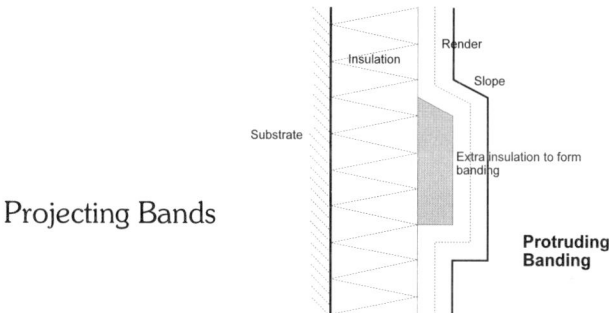

Projecting Bands

Render
Insulation
Slope
Substrate
Extra insulation to form banding
Protruding Banding

The projecting bands are best formed using addition insulation pieces shaped to the desired shape and size. Render is applied in the usual manner.

As a cautionary note, ledges in render should slope outwards and downwards to shed moisture avoiding undesirable mould growth.

Window Band

Usually projecting, these bands are a favourite feature in "classical" style elevations. The projecting "window or door band" should project in front of the adjoining wall finishes, accuracy and sharpness is essential to deliver a clean line to external arises.

These features are usually created with the use of timber battens, temporarily fixed to the basecoat. These enable the band feature to be built-up as necessary and finished prior to the final application of the adjoining wall finish.

Ashlar Cuts

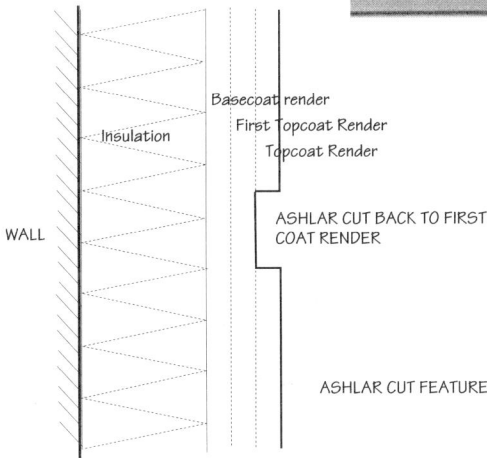

Basecoat render
First Topcoat Render
Topcoat Render

Insulation

WALL

ASHLAR CUT BACK TO FIRST
COAT RENDER

ASHLAR CUT FEATURE

The design of the cut, i.e. the size and depth of cut is dependent on the type and thickness of the render. The simple ashlar cut is created with three layers of render, one basecoat and two topcoats. The cut is made through the topcoat, back to the middle or first topcoat whilst the render is still uncured but firm, thus allowing the workman working time to create the desired effect.

Tile Hanging   Tile hanging is a traditional method of cladding walls, with many buildings using this finish variation to embellish and enhance featured elevations. The method is to hang tiles off timber battens which are in turn fixed to vertical timber bearers secured to the structural wall.

This simple system allows insulation to be inserted between the vertical bearers, insulating what has previously been a cold structural wall.

Typical Tile Hanging Installation

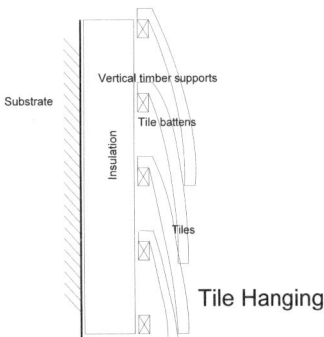

Tiles and other finishes

Suitable airways are required between the insulation layer and the sarking felt to allow the cladding to breath and avoid condensation forming.

Interesting architectural effects can be achieved with compatible adjacent finishes providing elevational variations to the "street scene".

4.10 TERRACOTTA Tiles
Featured System by Telling (Architectural Terracotta) Limited

Eisenberg Terracotta facades are manufactured under the 'Argeton' process, a process developed by the German Clay Tile Manufacturers Association to broaden the use of their expertise into construction claddings.

Manufacture of the tiles is undertaken at a purpose built factory in the east of Germany by F.V. Mueller Dachzieglewerke, the third largest German clay tile manufacturers. The location of the plant was selected due to the excellence of clays from that region which under their state of the art firing process create a product of superior impact resistance and lower porosity than traditionally produced clay products.

The Eisenberg Argeton System is constructed to comply with DIN 18516 Part 1 relating to ventilated non load bearing structural claddings.

Designs are in accordance with BS 6399: Part 2: 1995, to withstand loadings specific to the geographic location of the individual project.

Eisenberg Argeton is designed without gaskets, pressure clips or seals that can deteriorate within the design life of the system, will not creep under thermal expansion or contraction and tile fastenings are mechanical to ensure their integrity in the case of fire.

The system can be used in both new build and refurbishment in traditional rainscreen form installed to concrete or blockwork

substrates or as part of a unitised curtain wall or composite panel construction.

Telling Architectural holds leading knowledge of lightweight steel walling systems stick built on site or prefabricated to receive the terracotta rainscreen.

The system has been evaluated in use in Germany on structures up to 100 metres in height.

Thermal Performance :

Mineral Fibre or Phenolic of an appropriate density and thickness to DIN 18165 Part 1 or Part L Building Regulations are recommended for use with the system.

This can be varied to provide exceptional performance according to the wall construction. The insulant is mechanically anchored to the structure within the ventilating cavity to prevent detachment which would inhibit airflow. A vapour permeable membrane or foil face can be specified to protect the insulant against damp.

Fasteners should be non corrosive or synthetic type with plate heads.

Insulants on timber or steel structures can be adhesively bonded.

A minimum ventilated cavity of 30mm is required under DIN Standards between the outer face of the insulant and the inner face of the tile within the cladding zone.

4.11 Mouldings

Manufactured from pre-formed concrete, these mouldings are available in a variety of profiles to suite differing design requirements.

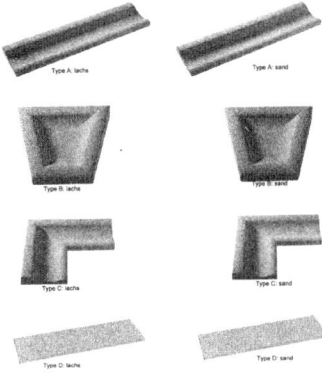

Surrounds for windows and doors can be enhanced or designed to comply with planning requirements such as the requirements for conservation areas or listed building consent.

Complete kits are available that include mouldings, adhesives and jointing compounds.

4.12 Block bonding

Block bonding effects can be achieved with the use of cementitious renders, suitably trowelled smooth with the designed block patterns cut into the surfaces.

The render has to be dry but not cured to enable the block pattern to be cut into the render surface without any detriment to the render.

Dimensions are established with caulk lines prior to cutting the features. The depth of cut is established (1-2mm) following to the application of render, but before it is cured, in order that the performance of the render is not impaired.

Note of caution, the deeper the cut the more likely the formation of "wet" marks or staining resulting in the formation of green mould, particularly in damp climates or north facing walls.

4.13 Specifications

Detailed specifications should be produced for every project where an external wall insulation system is to be installed. The specification is to include the complete instructions on installing the system including details of all trims, beads and any special features if required.

NBS Specification formats can be used where appropriate.

The following indicates those headings which should be expected in any project specification :-

1. **Title** - to describe the nature of the project in simple terms

2. **Contract Owner or Client** - to include all contract details including type of contract document to be used.

3. **Contact references** - to include all :-
    relevant addresses,
    telephone numbers,
    fax numbers,
    email addresses
    relevant websites.

4. **Description of project** - Brief particulars of the project to include :-
    Location of project
    Statement on insulation requirements
    Statement on finishes required.
    Any other matter of relevence.

5. **Contract References**
    References to be requested when/where necessary.

6. **Welfare Facilities** - Provision to be made for the welfare of all workmen, to include :-
    1. Toilet facilities
    2. Canteen facilities

7. **General Facilities** - Most contracts require the provision of the following facilities :-
   1. Electrical supply (110v)
   2. Water
   3. Storage for materials (Undercover and open)
   4. Office space

8. **Certification** - Statement on the use of certified systems, their type and suitability for the project. Note BBA/BRE Certification is recommended.

9. **Preparation Works** - to include the following :-
   External plumbing, removal or adjustments - due tolerances to be made for later installations generally but in particular at trim locations.
   External rain-water goods, removal or adjustments
   External signs, fixings, satellite dishes, fences, gates, and any other fixture or fitting secured to the external wall by any interested person that may have to be either removed or repositioned.

10. **Removal & Replacements** - Contracts can include the removal and replacement of windows and doors.

11. **Damage & Repairs** - Damage both externally and internally can be incurred, include references to cover such eventualities.

12. **Un-insulated Structures** - Contracts can involve attachment to, retention or exclusions of adjoining un-insulated structures. Suitable design allowances must be made to cover for such locations.

13. **Wall testing for fixings** - "Pull-out" tests to be carried out to enable the system supplier to certify the correct type and number of fixings to be used to ensure the system's security is guaranteed.

14. **Protection of the System** - Adequate protection for the system must be provided either by the specialist insulation contractor or main contractor. Protection must be from the following :-

    1. Weather conditions both extreme high and low temperatures.
    2. Impact damage from external sources.
    3. Damage from associated trades.
    4. Damage from vandalism.
    5. General protection of external areas adjoining.
    6. Protection of materials stored for use.

15. **Site Storage** - Adequate provisions to be made on site to protect the materials from damp, frost, vandalism and damage.

16. **Site Access** - Adequate access is to be provided by The Client, Main Contractor or Contract Supervisor to enable the system to be correctly installed without detriment to the process.

17. **Health and safety** - All current legislation to be complied with.

18. **Scaffold and site movement of materials** - Scaffold and any other so described access facility must comply with Health and Safety standards but at the same time provide enough work space to enable the workmen to install the system correctly. This usually provides for free-standing scaffold sited approximately 300mm from the wall surfaces.

    Notes :-
    1. Where scaffold has to be restrained through the system, due allowance and suitable design of the restraint is necessary.
    2. Scaffold should be free of internal braces to allow the installing contractor freedom of operation.
    3. Where work is to a multiple number of properties, state the number and type of properties to be scaffolded at

any one time, together with any contractual planning implications to completion targets.

19. **Thermal performance required to be achieved** - The thermal performance of the system when installed and completed is to achieve current Building Regulation Standards. Any deviation from this should be agreed with the client and local Building Inspectors.

20. **Finishes** - The final finishes are a clear requirement to be correctly stated with the following :-
    Colours
    Specified Aggregates, size and type
    Texture coating designs, textures and aggregate sizes
    Brick slip colours, size and types and suppliers
    Brick slip joint colours
    Simulated brick colours
    Simulated brick joint colours
    Simulated stone colours
    Simulated stone joint colours
    Dry cladding type, colour and sizes
    Any other type of cladding, colour size and type

21. **Samples** - Sample walls on site to be constructed for all interested persons to agree standards and colours prior to commencement of work.

22. **Mastics and Seals** - Suitable seals and mastics to be specified to suite local conditions.

23. **Flashings and Cappings** - Details of all flashings and cappings to be included, including the colour and type of finish and method of fixing, sealing and expansion joints where necessary..

24. **Copings** - Full detailed drawings to be provided to ensure adequate performance together with colour and type of finish and method of fixing, sealing and expansion joints where necessary..

25. **Accessories** - Accessories such as cills, window pods and special features to be fully detailed and specified.

26. **Refix clients accessories** - Refix such items as satellite dishes, fencing etc..

27. **Refix external plumbing, rainwater goods** - Clear definition to be made between specialist items and main contract items.

28. **Refix any other item as may be required**. - Clear definition to be made between specialist items and main contract items.

29. **Phasing** - Where work is to be phased, details of the phasings planned.

30. **Site clearance** - Areas of responsibility as to who is liable for site clearance to be indicated.

31. **Guarantees and completion certificates** - To be issued on completion.

32. **Remedial works - Guarantee Period by Specialist System Installer** - All to be clearly stated in the specification document.

33. **Payment details** - To be clearly stated in the documentation prior to tendering to include all retentions and periods of retention.

4.14 New-Build Systems

The requirement for highly insulated buildings providing energy efficiency, internal comfort levels and good economics, can be achieved by the use of external wall insulation installed onto a variety of structures. These are principally :-

1.   Mass concrete blocks and/or in-situ concrete.
2.   Timber-framed constructions.
3.   Insulated permanent formwork.

The first option of concrete blocks or in-situ concrete is probably the best for thermal storage, but the other options can easily achieve the requirement for complying with the current Building Regulation (Thermal Insulation) standards.

Mass Concrete or Masonry   The use of dense concrete blocks for "New-Build" combined with thermal storage and closed ventilation systems is greatly assisted by external wall insulation. The dense concrete aiding the thermal storage capacity and acting as a "Heat-Sink" or a thermal storage facility.

At construction stage, the use of an external wall insulation system is not so invasive as for refurbishment projects as all design considerations with windows and doors can be pre-planned.

The general detailing, whilst may be different to a refurbishment project is more conventional and the more costly elements of refurbishment, such as below ground floor level, can, to some extent, be designed-out.

Where this form of construction is used, i.e. solid wall - externally insulated, the attention to detail is essential to

1.   Avoid any "cold-bridging"

2.   Be weather-proof

3. Avoid total reliance on mastics to seal the system as there is the ultimate possibility of water ingress through the solid wall. *(If blocks are used)*

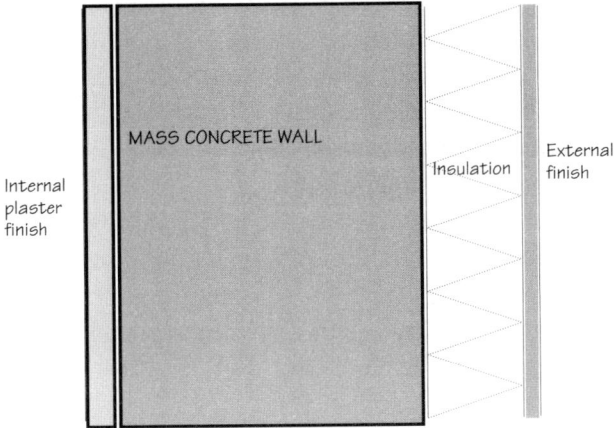

TYPICAL MASS CONCRETE "NEW-BUILD" WALL

EWIS Covering new-build construction

Pre-cast concrete panel constructions
Featured System by Eurobrick

Assembly by factory made panels gives for greater control over quality and delivery and helps eliminate problems due to inclement weather.

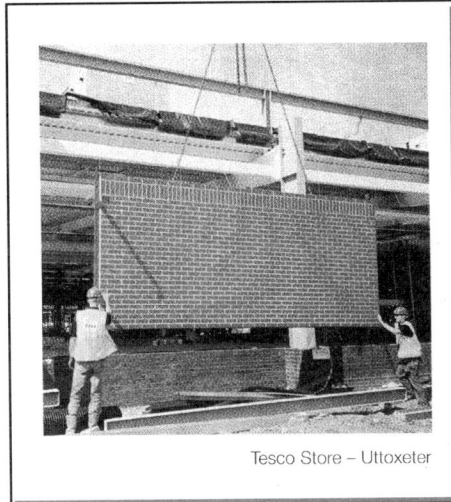

Tesco Store – Uttoxeter

Panel by Eurobrick Limited

Modular volumetric construction is already well established in the UK and Europe and following the Egan Report, "Rethinking Construction" more and more sectors are looking at off-site prefabrication as the sensible, affordable route to high quality buildings finished on time and with minimum disruption.

Assembling components in this fashion gives far greater control over quality and delivery schedules and eliminates the traditional construction problems of site hold ups due to inclement weather.

Eurobrick wall panels can also be factory produced in panel format for larger buildings. Complete walls from brick face through insulated frames to dry lining on the inside can be delivered to site and bolted together in a fraction of the time of "wet masonry" construction methods.

Insulated Permanent Formwork Systems   Increases in thermal performance requirements by the recent Building Regulation changes have resulted in the system of insulated permanent formwork systems becoming, at long last, economical to use.   These systems have been included as they relate well to the detailing requirements of external wall insulation systems.

Several companies now offer similar systems of moulded polystyrene blocks, assembled to form shuttering for concrete. Known now as **ICF** Systems, these are illustrated in brief as follows :-

Beco**Wallform**
Insulating Concrete Formwork (ICF) System

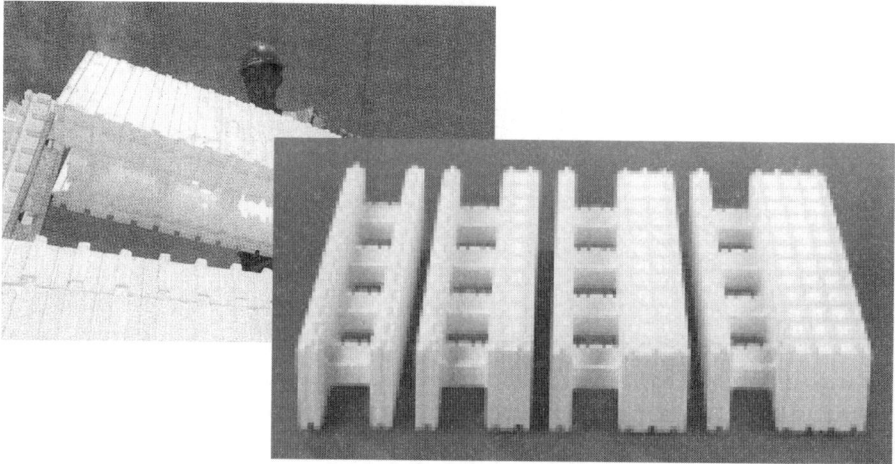

An insulating concrete formwork (ICF) system, Beco WALLFORM is based on large hollow lightweight block components that lock together without intermediate bedding materials to provide a formwork system into which concrete is poured.

Once set, the concrete becomes a high strength structure and the formwork remains in place as thermal insulation, with u-values ranging from 0.30 W/m²k down to 0.11 W/m²k - ideal for zero energy buildings. The building process is quick, tidy and precise, with lower labour and equipment requirements than alternative methods.

198

PART FIVE - Installation & Repairs

5.1   Wall Preparations

Walls are constructed with a wide variety of materials and require differing methods of preparation, dependent on site examination and conditions. Most walls can be brushed-down, washed over with clean water and treated with a fungicidal wash as necessary. Testing must be carried out to determine the correct type and quantities of mechanical fixings.

The vertical alignment and regularity of the surface of the substrate must be checked with a straight-edge. The areas of point of contact are particularly important for adhesive based systems. Dubbing-out may be required to correct uneven surfaces to ensure good areas of points of contact.

Where adhesives are planned to be used particular attention must be given to the surface to remove completely, paint, contamination of any type or old coatings to ensure that a sufficient bond is achieved.

All areas where trims and special accessories are to be installed are to be checked for an even surface and dubbed-out level if required. Without this attention unsightly distortion of the trim may occur.

Cleaning & Testing The substrates may comprise any or all of the following:-

1.   Brickwork and blockwork

2.   Painted masonry surfaces

3.   Concrete

4.   Old Render

5.   Natural Stone

6.   Timber or cementitious boards

199

All wall surfaces must be examined and if necessary prepared in the appropriate manner for the application of the system, to be used in accordance with the following:-

Brickwork & Block-work   Usually reasonably flat, it should be examined for soundness, surface contamination of mould growth, de-lamination of the surface by frost attack and any voids that may be present. Any surface contamination should be removed and cleaned. Where mould growth has been present treatment of a suitable fungicide is to be applied. Severe undulations should be dubbed-out with a sand/cement mortar, bonding agents such as PVA or SBR may be necessary.

Painted Masonry Surfaces   May be difficult to adhere-to with adhesives and renders without any preparation such as scarifying old paintwork which should be removed by approximately 70%. However, mechanical fixings can penetrate the surfaces by drilling to enable conventional fixings to be used. The type and thickness of the paintwork must be considered to determine if there is any problem with potential condensation risks.

Concrete   Concrete surfaces may be available in differing textures and finishes dependent on density, weathering and dilapidation. Test for density , remove any growth as necessity and examine for levelness of surface. Any exposed reinforcement should be treated for rust and suitably pointed.

Old Render   Old render should be examined for de-lamination by hammer test. Any areas found hollow to be removed and the areas "dubbed-out" with a sand/cement mix leaving the area continuous with adjoining finishes. All other recommendations as for brickwork and block-work also apply.

Natural Stone   The very nature of natural stone creates an uneven surface of texture and hardness, joints may be porous, close examination is essential where adhesives are to be used, dubbing-out is usually required, if in doubt, render flat the whole wall surface. Where mechanical fixings are planned, any uneven surface can accept to a degree slab insulants of a reasonable thickness, however, some dubbing-out still may be required dependant on circumstances. Treat with a proprietary fungicide solution.

New-build dry sheathing   New pre-fabricated structures may be clad with ply or cementitious boards, which may have been installed on site or in the factory prior to delivery. Before cladding commences, the sheathing boards must be checked for satisfactory fixings to the supporting framework.

All joints to be sealed and suitable protection installed if the sheathing is to be exposed to the weather for considerable periods. Impact protection may also be appropriate on exposed corners or impact vulnerable areas.

5.2   Methods of cleaning & Testing

Walls generally are to be clean and free from deleterious material prior to the installation of an external wall insulation system, particularly if adhesives are to be used as the principal method of fixing. Various methods can be adopted, dependent on the condition of the wall surface. A small test area may be appropriate for any system to avoid the possibility of excessive or undesirable reactions.

Water Cleaning   Cleaning by "power-washing" uses high pressure spray equipment, steam can also be used where necessary. Approved detergents, cleaners and caustics can be incorporated in the process. The substrate should be allowed to dry before the application of any system.

Mechanical Cleaning   Hand cleaning using power-tools, such as grinders, sanding disks or rotary wire brushes are suitable for some small areas, however, for large areas dry or wet grit-blasting methods may be more suitable. Grit blasted surfaces should be hosed-down on completion and allowed to dry prior to the application of any system.

Chemical Cleaning Acids and solvents can be used to dissolve unwanted coatings, sealers or paint. Wet all surfaces well prior to the application of any treatment and complete the process by a thorough wash-down of clean water. Allow to dry before application of any system.

The suitability of water, mechanical or chemical cleaning methods varies with substrate type. Masonry and concrete tend to be porous, damp and highly alkaline. They can be damaged by an incorrect cleaning process. It is recommended that a small test area be treated prior to any cleaning application to determine effectiveness.

5.3 Methods of selecting Fixings

The substrate must be sufficiently stable and robust to support the external wall insulation system, but testing may be required on differing areas of particular projects to ensure continuity of standards.

Hammer testing may be required to identify areas of wall which may "bulge" or sound hollow. These areas may not support adhesion or have insufficient strength to comply with the requirements of a "Pull-out" load test.

After the cleaning procedure has been completed, together with any dubbing-out of the surface, "pull-out" testing is required to establish the failure load of all of the proposed fixings. If there are many substrates, all must be tested. As a result of these tests, the final design layout of the fixing pattern is concluded with a fixing plan to be issued to the site operatives and contract supervisors.

There are many strengths in differing substrates, with always a possibility of voids. Variable qualities in mortar, joints and rendered surfaces effectively hiding the quality of structural elements, making the "pull-out" test an essential procedure.

No-Fines concrete will produce a wide range of test results due to the nature of the no-fines walling.

Insulating concrete blocks, whilst very good in their thermal performance, are very friable and weak as far as strength to resist "pull-out" loads are concerned requiring special fixings or an increase in quantity of fixings.

Stonework can vary significantly with the type of stone encountered. (sandstone to granite).

Examples of substrate variations are as follows:-

Aerated blocks	2.8 - 7N/mm
Aggregate blocks	3.5 - 10.5N/mm
Dense aggregate blocks	7 - 24N/mm

(The values stated are an indication of the material strength)

The procedure for the "pull-out" test is generally in accordance with the instruction of the manufacturer's equipment, but the following are guide-lines:-

a. Visually inspect the wall or substrate and determine suitable areas for test which should be accessible, flat and free from loose material or contamination.

b. Check by hammer test if there is any de-lamination or hollows to the surface in the test area

c.   Drill the correct diameter holes, in several locations to the correct depth using a suitable power drill fitted with a suitable depth gauge.

d.   Hammer into the hole the fixing to the required depth.

e.   Fit the "Pull-out" metre to the fixing head and operate the metre to withdraw the fixing, under load, and read the gauge.

f.   Record all readings of withdrawal under load and note, address, location, date, time, weather conditions, type of substrate and any other matter likely to be of importance to the specifier.

Fixing manufacturers and suppliers together with system promoters offer the service of carrying-out the "pull-out" operation and report the results accordingly. From these values, safe-working loads taking account of wind-loadings can be assessed.

Factors to be considered for fixing selection:-

1.   CP3 to be complied with - Wind load (Highly variable)

2.   Total dead load of system including finishes

3.   Bending load of heavyweight finishes

4.   Fire resistence

5.   Material of sub-strate.

6.   Practical installation methods

Testing record sheet available on CD ROM disk

Selection of Illustrations

Setting-Out Brick Render

Finished "Brick Rend" Effect

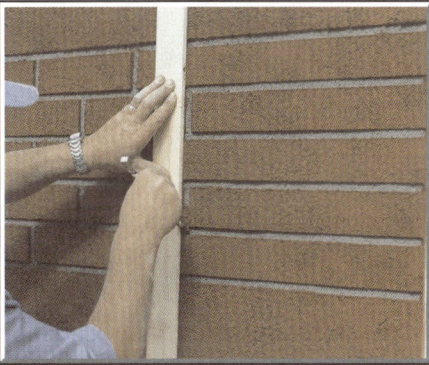
Setting Verticals

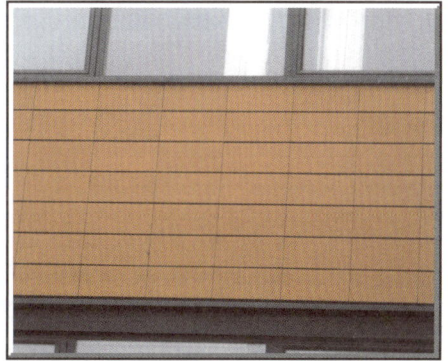
Terra Cotta Tiles

Brick Slips on Mesh

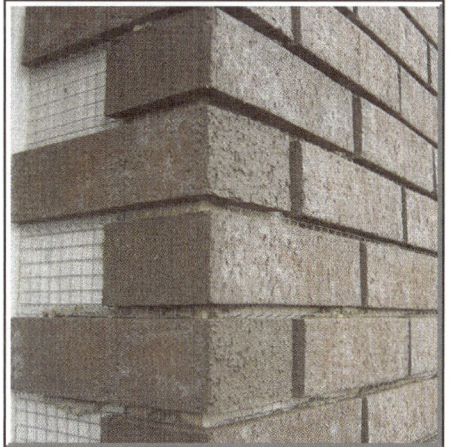
Brick Slip Corner

Selection of Finishes

Block Bonding in Render

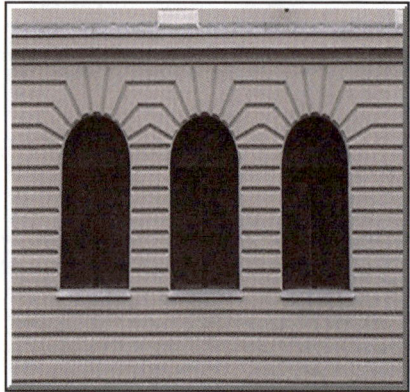

Ashlar Effects

Spray Render - Bathstone

Spray Render - Terra Cotta

Some Brick Slip Colours

Brick Slips & Porch

Selection of Applications

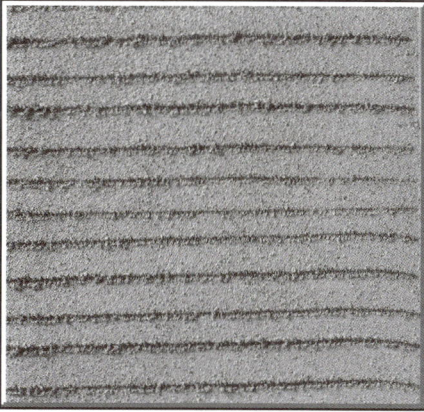

Basecoat

Dashing Coat

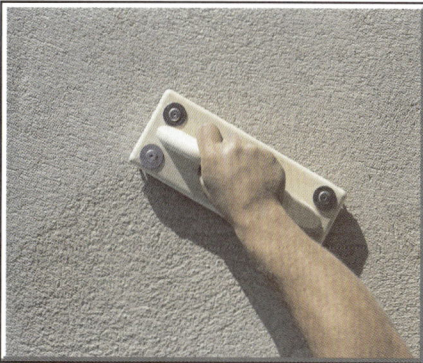

Scratching Process

Scratch Finish

Roughcast Application

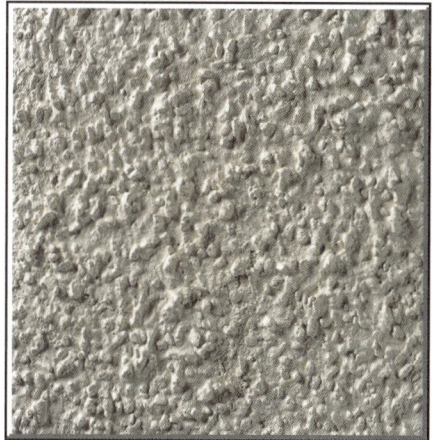

Roughcast Finish

Selection of Colours for Finishes

Terracotta Render

Buttermilk Render

Cream Render

Sandstone Render

Oatmeal Render

Polar White Render

5.4   Pre-Installation Preparation

A pre-installation and pre-design procedure for installing insulation boards is advisable, the following to be checked prior to any installation:-

1. Hammer-test all wall surfaces to determine any hollows or de-lamination.

2. Repair wall as necessary following the hammer-test.

3. Treat any walls surfaces with growth contamination with a suitable proprietary fungicidal wash as required. Dilution of wash to suit circumstances and to follow manufacturers instructions.

4. Brush wall surfaces with a stiff brush to remove residual dead growth matter.

5. Remove any unnecessary features, signs, protrusions, and fitments to enable the insulation work to proceed.

6. Make good any defective areas and apply suitable stabilizing solution if necessary.

7. Use high adhesive polymer modified mortars as required.

8. Any temporary plumbing or rainwater goods to be re-sited as required before work proceeds.

9. All builders work to be completed before work proceeds.

Notes on drilling and pinning
1. Always use a depth gauge and drill level.

2. Always ream holes clear of dust before pinning.

3. If type of fixing or strength of substrate is in doubt, carry out a full "pull-out" test.

4. Check old brickwork, blockwork etc.. continually as use of percussion drills can damage or dislodge substrates.

5. Work out fixing pattern per sq m as soon as possible.

6.   Do not overdrive fixings.

7.   Do not distort fixings.

8.   Always insert fixings straight, not at any angle.

5.5   Pre-Installation of Beads & Trims

A pre-installation procedure for installing beads and trims is advisable, the following to be checked prior to any installation:-

1.   Check wall/substrate for alignment and levelness, small tolerances may be permit-able but subject to the trims not distorting.

2.   Remove any protrusions, wall imperfections as necessary

3.   Dub-out surfaces as necessary to achieve flat workable surfaces

4.   Repoint damp-proof courses if required *(they may be hidden from future examination after the wall is insulated)*.

5.   All trims should be installed in as long a length as possible with joints kept to a minimum.

6.   Any overlapping joints shall be a minimum of 50mm

7.   Protective tapes shall only be removed on completion of the works wherever possible.

8.   Attention is drawn to "line and level" particularly for those trims at "eye" level, but equally important for all trims.

Trims and accessories can be divided into two categories, the first being those trims and accessories fixed directly to the substrate, the second being those beads and accessories fixed on the face of the insulation or finish.

Installation of Beads & Trims Trims and accessories are usually supplied in coated galvanised steel, stainless steel or uPVC. and comprise Base trims, End trims, Top cappings, Structural movement joints and window cill extensions (over & undercills). They are usually supplied in lengths of 2500mm & 3000mm. Overcills are usually specially manufactured to fit specific openings on a "bespoke" basis.

Trim fixings should be compatible with the trim material to avoid any chemical reaction resulting in corrosion. Window cill extensions are usually manufactured in aluminium and completed with a coloured powder-coated finish. Trims can be supplied with or without a coloured uPVC nosing, this is optional and will depend on the designers requirements.

All substrate surfaces must be level and without undulation to ensure full even support and fixing, failure to achieve reasonable flatness will result in trims distorting along its length. Fixings should be installed at a maximum of 600mm centres using fixings suitable for the substrate to which the trim is fixed. Joints should be lapped by a minimum of 50mm except for exposed feature trims and top cappings, which will be provided with suitable jointing plates.

Corners, particularly for the base trim should be pre-manufactured with a welded joint.

Profiled trims and features are an alternative to edge-rendering and offers several advantages such as,

They are relatively easy to install, especially at straight system terminations.

They provide a truer, straighter edge than a rendered edge

They can provide coloured design additions to any architectural feature

211

They are specially manufactured to suit any insulation or render thickness

Where trims and accessories are used the following procedures should be adopted wherever possible:-

1. Design or select the correct trim necessary to perform the required function for the location.

2. Select the correct colour, nosing or profile

3. Ensure joints have the correct jointing plates, sleeves or seals

4. Ensure trims are fixed to the substrate with the correct fixings and at the correct centres.

5. Ensure any overhangs are sufficient to shed water away from the system

6. Ensure the trims are sealed correctly to the substrate with suitable mastics.

7. Only use trims manufactured with a suitable durable material and painted/coated with a appropriate certifiable coating.

A selection of detailed trim drawings for typical project locations can be found in part four of this manual.

Fixings provided to install trims and accessories shall be suitable for the purpose and compatible with the material to be secured, i.e. avoid mixing differing metals in direct contact. Generally all fixings should be installed at 600mm centres using the fixing holes provided by the manufacturer. Fixings should penetrate into masonry by 40mm, where they are fixed to a board substrate, the fixing should be of a screw type with the threads fully locked into the board thickness.

Joints in trims should be provided with suitable jointing plates to allow proper sealing with mastic. Screws should be inserted through into the joint plates to prevent failure of the joint.

All trims, flashings and cills shall be weather sealed with a suitable mastic between the fixing flange of the trim and the substrate. Welded junctions, corners and transitions should be used wherever possible.

All shall be cleaned to remove any surface deposits that may cause corrosion or deterioration of metal surfaces, cills should be provided with "low-tack" protective film to be removed on completion of the work.

Bellcast - base-trim Bellcast base trim shall be fixed, true to line and level, in full lengths, wherever possible and shall be overlapped fully at all corners and at each straight joint. It shall be mechanically fixed, after drilling through the back face with fixing pins, at 300mm centres with additional pins through each joint. Where necessary provide pins 50 mm from each end.

Base-trim fabricated corners Corner base-trims shall be fixed true to line and level in the same manner as the straight lengths.

Corner beads Corner beads shall be fixed, true to line and level, after fixing of insulant and before the first coat render by either of the three methods listed below:

1. By applying Bedding Mortar at maximum 300 mm centres on both sides of the corner into which the bead shall be pressed. It shall be in full lengths wherever possible and shall be butt jointed. The bedding mortar shall be allowed to set before applying the base coat mortar to the general areas .

2. Wiring to the galvanised/stainless steel mesh.

3. Direct fixing to the substrate through the insulation board.

213

Top Trims The top trim shall be fixed, true to line and level, by applying silicone sealant to the back face of the profile and pressing the trim into position. Prior to fixing the insulant, ensure that the sealant is distributed to form a watertight bed and the trim is fixed securely.

The stop trim shall be fixed in full lengths, wherever possible, and shall be butt jointed, mechanically fixed through the back face, with profile fixing pins, at 600mm centres, with additional pins 50mm either side of a joint and 50mm from each end. Jointing sleeves to be installed as required to ensure weather-tightness.

The jointing sleeves shall be fixed by applying silicone sealant to half of the sleeve and pushing it into the stop trim section. Silicone sealant shall then be applied to the other half of the clip and the joining trim section pushed onto the clip. Any excess sealant should be removed from the joint.

Surface stop beads   Surface stop beads shall be fixed generally vertically, true to line and level, after fixing the insulant and before the base coat render by one of the methods listed blow: -

A	Apply Bedding Mortar at maximum 300mm centres into which the bead shall be pressed. The Bedding Mortar shall be allowed to set before applying the base coat mortar to the general areas.
B	Direct fixing through the insulant.
C	Wired to a metal render reinforcement mesh.

Surface movement joints   Surface movement joints shall be fixed vertically only, true to line and level, after fixing the insulant, and before the first coat render. The insulant shall have a continuous full depth cut through it on the line of the joint. The width of the cut shall not be less than 2mm and not greater than 5mm.

The joint shall be fixed by offering the joint to the insulant with the nosing over the cut. Insert surface bead retainers through the baseplate of the bead. Secure with retainers at maximum 300mm centres and extra retainers at no more than 50mm from each end. The beads shall be in full lengths, wherever possible, and shall be butt jointed with the plastic nosing extended so that the joint in the nosing is covering 25mm over the joint in the metal work.

Any excess nosing, at the end of the movement joint, shall be trimmed flush with the metal work using a sharp knife. uPVC movement beads are butt jointed.

Structural movement joints   Full-depth movement joints are used to continue the structural construction joint through the external wall insulation system. The trim is fixed directly to the substrate either side of the existing joint. Due to variations in movement joints used in the building substrate and the likely possible high degrees of movement, due care must be taken to choose the correct sized (width) trim.

Sufficient fixing area must be allowed either side of the structural joint and fixings must be well clear of the existing sealants. If there is an alignment variation in the existing structural joint this may prove difficult to straighten visually with the new applied joint. The structural joint may take one of two forms namely:-

a.   Joined with a uPVC flexible extrusion

b.   Joined with an applied flexible mastic

The type "a" is relatively simple to install with the extruded uPVC sealing both sides of the joint, this extrusion being designed for its purpose and is fully weather-proof.

Type "b" joint requires on site application of applied mastics and will rely on workmanship to fully protect the joint.

Aluminium flashings/accessories Aluminium powder-coated trims are used primarily to weatherproof difficult locations on the face or top of a building element and are manufactured specially on specific jobs.

Joints designed within the trim and any welded corners must be sealed with silicon mastic.

5.6   Installation - Insulation boards

Installing insulation boards or slabs whether by adhesive or mechanical fixings should always follow procedures to provide a flat even surface, with the minimum joint gaps.

Damaged boards or badly cut boards should be discarded ready to be cut down in size for those smaller areas requiring measured cut pieces.

Where distorted joints occur, boards badly cut or any other reason for the insulation to be dis-continuous, foam should be injected into the gap or space to ensure insulation continuity. Surplus foam to be removed prior to the application of any finsh.

Adhesives Adhesives to secure insulation boards are always applied to the insulation board surface and not to the substrate. Once coated, the boards are located in the required position and pressed lightly with even pressure as soon as possible after spreading and before the adhesive commences setting.

There are two basic methods of applying adhesive, a notched-trowel method or ribbon-and-dab method. The former is best for flatter surfaces, the latter for application on uneven surfaces.

The notched trowel method is to lay an even spread of adhesive over the whole surface of the board. The ribbon method consists of applying a ribbon of adhesive around the perimeter with dabs on the interior.

The ribbon and dab method will require greater quantities of adhesive than with the notched-trowel method.

Whichever method is adopted, the quantity and distribution of the adhesive is important both for security and cost, sufficient adhesive must be applied to be secure, however, too much adhesive is wasteful and costly. The "area-of-point-of-contact" is the ultimate test of which method should be adopted as the greater this area the greater the security of the installation.

A small test is recommended on each varying substrate surface to evaluate the "area of point of contact" for the system used. If this area is insufficient then the wall surface may, by necessity, have to be rendered to correct surface undulations. Once applied, the board should not be removed as the adhesive will probably be cured or be destroyed and made further unusable.

Edges of boards should be kept free of adhesive to avoid gaps being produced in the insulation boarding thus creating cold-bridges. Some systems and locations may require edge-wrapping of the reinforcement layer at perimeters.

Direct Mechanical Fixings   Mechanical fixings can be used to fix the insulation boards directly to the substrate.

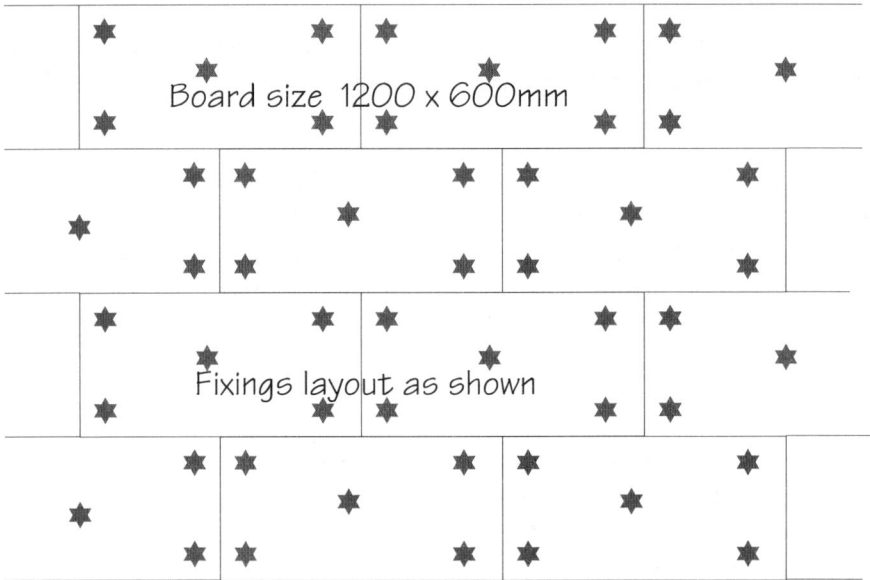

Board size 1200 x 600mm

Fixings layout as shown

Fixing insulation boards directly to the substrate requires a fixing pattern that relates to the board size and layout,

Where there is a requirement to fix through the mesh and basecoat together with the insulation board, a grid of fixings to provide an approximate fixing arrangement of 8-10 fixings per m^2 is desirable.

A grid of 400 x 300mm provides a rate of 8.3no per m^2 which is generally acceptable for most applications.

Rail Systems Some external wall insulation systems are fixed to the substrate by means of pvc or galvanised metal rails.

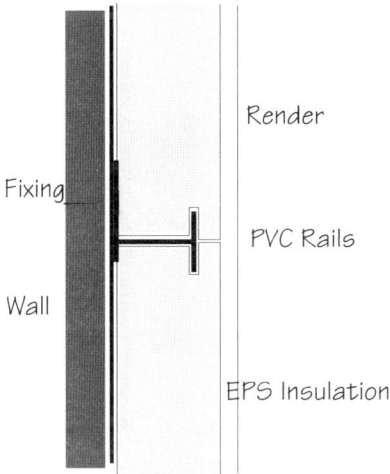

System being installed

The rails are fixed directly to the substrate with the interlocking flanges located within the insulation. It can be noted that expanded polystyrene is the most suitable insulant to be grooved in this manner as a greater thickness of insulation can be provided with this material than with other insulations. The layout drawing shows the principal of such interlocking systems.

This rail system is best used where walls are not continuous or where thick insulation is required either to level uneven walls or to achieve high standards of insulation. Expanded polystyrene is also the best insulation materials for factory shaping with the application of grooves, enabling the rails to be fitted.

Other insulants are difficult to shape in this manner. This system will rely entirely, in most cases, on the render finish coat being adhesively attached to the insulant, a mixture of rail fixing and combined with the use of mechanical fixings may be desirable on certain projects.

Preparing and Installing Insulation Boards   Insulation boards are lightweight "workable" materials that are easily cut to create customised shapes, profiles or ornamentation. Board cutting is a routine but critical aspect of board installation. Careful planning and layout using full sized boards wherever possible minimises the need for cutting.

Continuous vertical joints should be avoided wherever possible, interlock or block-bond as far as practical, particularly at corners. If a project has a large quantity of special sizes these may be factory produced as necessary.

Clean, sharp edges are required for proper board installation. Avoid wherever possible the use of damaged boards or boards with damaged edges which may create gaps or voids within the board joints. The inclusion of any gap or void creates a "cold-bridge" through the insulation and is undesirable. Any gap or void so formed should be filled with expanding foam and cleaned off to a smooth board face.

Precise board edges can be obtained by careful cutting with conventional hand tools or powered table saws. Electric hot-wire cutters are only suitable for expanded or extruded polystyrene.

Procedure for installing Insulation Boards   Proper installation of insulation boards should achieve a continuous thermal barrier that is able to accept subsequent applications of mesh reinforcement and finishes.   Installed boards must provide an essentially seamless surface with proper detailing at system terminations.   Although board materials and attachment methods differ by system type the following procedure is a good guideline.

1.   Establish the base line

Where a base trim is required, usually at damp-proof-course (dpc) level, but also installed at higher levels to create break-lines in finishes. This trim is to be installed first to a level line using a builders level, or other mechanical/technical device, commence boarding with the board on top of this trim.

Note, an existing dpc may not be level and suitable adjustments may be necessary. Where there is no base trim, install a timber batten to a level line using a builders level and commence installation on top of this temporary support.

2.   Establish vertical stops and any structural joints

Install all vertical full system stop trims and any vertical structural movement joints prior to the installation of insulation boards, any horizontal "through-system" trims can be installed as the boarding progresses.

3   Boarding

Adhesive fixed boards are installed either using the ribbon and dab method or the notched-trowel method. Ensure that all edges are cleaned of any adhesive as under no circumstances should adhesive compound be allowed to fill board joints. Butt-joint tightly all boards and apply reasonable constant pressure for uniform adhesive contact and high initial bond.

Avoid the use of hammers to straighten boards as this will cause damage to the board surface. Set a straight edge across several boards in all directions to check that boards are level and flush.

Remove the occasional board, whilst still wet, to check adhesive distribution and area of point of contact. Boards should be "block-bonded" in courses to avoid continuous board joints. Interlock boards round all inside and outside corners. Cut "L or U" shaped boards as far as it is possible when installing around windows and doors.

If gaps are created or are unavoidable, these should be filled with injected foam insulation and cleaned to a flush finish.

4   Insulation face adjustments

Where a lightweight render is to be applied, the surface of the insulation board is required to be flush and level without any joint displacements. The finish as applied will only have a thickness of 2-3mm and will show any irregularities left in the board surface left prior to the application of the render and finish.

Adjustments may be required to flush finish the board surface, note, only expanded polystyrene is suitable for rasping as other insulation materials will usually have a skin of glass tissue or have a fibrous constituency and therefore are impossible to trim by such means.

5   Edge-wrapping

Where metal or uPVC trims are not required or used, edge-wrapping with the reinforcement mesh is needed to cover the insulation and terminate the system at vertical locations. Edge-wrapping is to continue the reinforcement mesh around the board edge and behind the insulation board, in contact with the substrate for approximately 50mm.

The best method of edge-wrapping insulation boards is to pre-bed a strip of the mesh, approximately 250mm wide, dependent on insulation thickness, at the required locations, onto the substrate, after the boarding as been installed the mesh is bedded into the basecoat mortar on the edges and onto the face of the insulation by approximately 100mm. The surface reinforced layer is then applied, overlapping the edge-wrapping ensuring continuity of reinforcement.

Edge-wrapping with metal mesh must be carried-out with pre-bent mesh. The metal mesh is pre-formed into a "U" shape covering the back and front insulation faces by a minimum of 50mm. The face reinforcement covers the edge pieces and can be secured through both layers into the substrate.

5.7   Reinforcements

All renders are required to be reinforced, either by a lightweight mesh or a heavyweight metal mesh dependent on the design of the render. The final choice of reinforcement type relies on site requirements and the selection by the Architect or Contract Supervisor. Considerations such as vandal resistance, fire resistance and type of finish should also be included in the design.

These differing meshes have quite different physical properties and require differing renders and installation methods.

Lightweight Reinforced Renders   Reinforcing mineral fibre mesh is supplied in roll form, usually 50x1m and wherever practical, is cut into full height lengths. The mesh is placed onto the wall at the top and with a vertical motion downwards trowelled into the base coat mortar immediately after it has been applied. This base coat is usually applied at the rate of 2kg per mm thick per square metre.

The fixing pins are inserted whilst the mortar is still uncured, where required. The mesh shall be placed around all corners and into all reveals and have minimum laps of l00mm at all joints. The mesh shall be hung downwards and cut flush with the bottom of the drip of the bellcast. The mortar which penetrates through the mesh shall be trowelled smooth with additional mortar to fully cover the mesh.

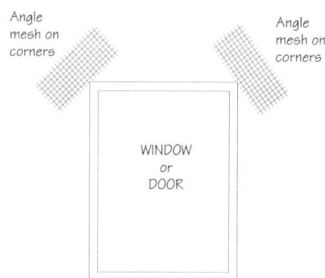

Angle mesh on corners

Angle mesh on corners

WINDOW or DOOR

Additional strips of reinforcing mesh 500mm long by 250mm wide minimum shall be applied length ways across the corners of all windows, doorways and other openings such that they extend equally either side of the corner.

The corner reinforcement strips shall be applied after application of the wall reinforcing mesh or armoured mesh. Detailed mesh is applied to external corners and architectural features as required.

Installation of mechanical fixings   Masonry fixing pins shall be inserted into 8-10mm diameter drilled holes with 50mm or 60mm minimum wall embedment dependent upon fixing type. The fixing holes are drilled through the basecoat render *(whilst still uncured)*, reinforcement, insulation and into the structure. They shall be hammered home until the heads of the pins are flush with the surface of the base coat before the centre pins are hammered home.

Any excess mortar shall be smoothed over with additional mortar being applied as necessary, the final finish being lightly scratched horizontally.

The fixing pins shall be positioned on a square grid at a maximum spacing of 300/400 mm horizontally and vertically. Additional pins shall be inserted, if necessary , to ensure that the maximum distance between the fixings shall not exceed 300mm. Fixings adjacent to external corner and rows shall not exceed 150mm.

All lightweight glass fibre meshes must be installed correctly in order that they reinforce the basecoat to provide, water resistance, minimise cracking and improve impact resistance. The glass or mineral fibre mesh is comprised of individual fibre strands woven into a square mesh in a form to resist deformation. Suitable overlaps must be provided where necessary of approximately 100mm, additional diagonal reinforcement strips to be provided at any location likely to create stress points.

All system terminations should be edge-wrapped or appropriately provided with metal edging accessories, to include movement joints and points where the system abuts dissimilar materials such as windows, doors, electrical boxes, water pipes or other fixtures.

Avoid damaging the mesh, trowels and floats can cut mesh if not carefully used. Never cut the reinforcement to compensate for wrinkles, folds or blisters, always trowel flat.

Application of basecoat It is important that basecoats should never be installed in temperatures below 5degC, with a falling thermometer or onto surfaces that are frozen. Applications should be applied when temperatures are rising from 5degC. and not installed in very hot conditions whereby the curing processes are artificially accelerated.

Inspect the insulation board surface, if the surface has been damaged , check for loose particles and dust which should be removed. Where insulation boards are used that cannot be adjusted, check for board displacement and dub-out as necessary with the basecoat material, allow to cure.

Pre-cut glass fibre reinforcement mesh to workable lengths using a utility knife. Allow for overlaps of 100mm. Only cut sufficient mesh to cover a suitable working area to enable full application in a reasonable working period.

Install manufacturers pre-formed corner mesh or beads using the basecoat material. Use a builders level to check vertical and horizontal alignments.

Using a stainless steel trowel, apply basecoat to the insulation surface of the insulation board to the appropriate thickness, usually 3-4mm. Immediately place the pre-cut mesh on the wall into the wet basecoat. Using firm, even pressure trowelling towards the mesh edges, forcing the mesh into the basecoat. Avoid wrinkles and blisters wherever possible.

Apply additional basecoat material, as necessary, to totally encapsulate the mesh, trowel smooth the basecoat to achieve a uniform, even surface with no visible sign of the mesh pattern. Skim over with the basecoat any exposed meshed areas to complete the base ready to receive the finish coat. Systems receiving a further render of basecoat material must be lightly scratched or trowelled smooth ready to receive final finish.

Lightweight systems usually have the insulation boards fully fixed, either with adhesive or a combination of adhesive and mechanical fixings, prior to the application of the basecoat.

Medium weight systems are usually fully mechanically fixed with approximately 50% of the fixings installed through the mesh.

Installation of Detail Mesh Special conditions may require additional layers of reinforcing fabric. For edge- wrapping cut the fabric 250mm - 300mm wide or use the precut detail mesh to a workable length.

Use reinforcing mesh at all openings, such as; doors, windows and air conditioner openings.

Double mesh corners and place mesh strips, a minimum 300mm long and 250mm wide, at a 45° angle to every exposed corner. Additional reinforcing may also be accomplished by using meshes of varying weights.

Heavyweight Reinforced Renders Reinforcing metal meshes are usually supplied in stainless steel, although galvanised mild steel can be used, in sheet or rolled form. The sheets are usually in sizes 2500 x 1250mm and the rolls 5m x 1m.

The metal mesh is installed with the diamond expansion holes in a horizontal mode, laps are usually two diamonds wide and fixed using plastic or stainless fixings at an approximate rate of 8-9 fixings per m². Additional fixings are installed at corners and overlaps of mesh.

The laps are fixed with the system fixings, where necessary wiring the laps may also be required.

Installation of mechanical fixings - metal mesh Masonry fixing pins shall be hammered into pre-drilled holes 8-10mm diameter, 50mm or 60mm minimum embedment into the wall is necessary dependent upon fixing type. The fixings are hammered into pre-drilled holes which have been drilled through the, reinforcement, insulant and into the structure. They shall be hammered home until the heads of the pins are flush with the surface of the reinforcement without deforming the mesh into the insulation board.

The fixing pins shall be positioned on a square grid at a maximum spacing of 300mm/400mm horizontally and vertically. Additional pins shall be inserted, if necessary, to ensure that the maximum distance between fixings shall not exceed 500mm. Fixings adjacent to the external corner and row do not exceed 150mm.

Notes to observe when installing reinforcements

a   Always fix metal mesh vertically.

b   Always form arrises before fixing on wall.

c   Always fix metal mesh with diamonds horizontal.

d   Always wear protective gloves when installing metal mesh.

e   Always start from a top corner .

f   Always lap mesh by a minimum of two diamonds in either direction.

g   To get a snug fit into a corner bend mesh beyond a right angle.

h   Form arrises on mesh, using a timber batten or mesh bending frame, before fixing to wall.

I   Lapping sheets must be tied by snipping strands behind and bending over strands of front sheet or using galvanised or stainless wire as appropriate twisted around mesh.

j   Bulging mesh can often be flattened by running the flat face of a hammer firmly and smartly down the mesh surface.

Performance Criteria of Render Carriers   Render carriers should be resistant to degradation by the action of water or moisture. Galvanised products should be used with care. Stainless steel products should be grade 304 or 430 to the required BS.

Generally in the form of a square or diamond shaped mesh, supplied in rolls or sheets. Render carriers should be selected according to the weight and wind suction loads they will be required to resist.

5.8   *Application of Renders*

The Code of Practice (BS 5262 : 1991) provides for the requirements and good practices of the application of cement-based external wet renders on all types of conventional backgrounds and should be followed where ever possible.

Cementitious render coats should never be installed in temperatures below 5degC, with a falling thermometer or onto surfaces that are frozen.   Applications should be applied when temperatures are rising from 5degC.

Care should be taken when the weather is hot or when there is a strong drying air passage over the site of render application.   Quick setting of renders may be detrimental to the curing processes and if in any doubt advice should be taken from the manufacturer or supplier.

Render Notes

Storage   Render sacks, even with hoods, are only shower-proof, and should be protected from the weather and damp storage conditions.

Shelf-Life   Shelf-life is normally approximately one year in dry conditions in original packaging.

Test Panel   It is recommended that a test panel (ideally 2 square metres) be produced for inspection, and the client, architect, main contractor, clerk of works etc. should not let work commence until completely satisfied with workmanship. Texture quality, colour, depth of coat, should all be assessed. This should not be part of the main contract and must be prepared well in advance to allow time for curing and approval procedures.

Workmanship   Renderers must become familiar with the product's water requirement, work ability characteristics, setting and hardening time which vary according to type, background, temperature, humidity and orientation.

Mixing - Polymer Renders   Use approx. (4 -5) litres of clean water, (as directed by the manufacturer or supplier,) per 25kg bag. A suitable measuring bucket ensures the correct quantity and accuracy every time.

Mix thoroughly it takes at least 10 minutes to dissolve the powder additives. Note how the semi-dry mix becomes creamy after several minutes. Mix will normally be slightly sticky.

Do not add anything to the mixer other than clean water.

Reinforcement Notes

Metal Mesh   All galvanised and stainless steel meshes must be installed correctly in order that they reinforce the basecoat to provide, water resistance, minimise cracking and improving impact resistance.

The metal mesh is comprised of either post-galvanised mild-steel expanded mesh or stainless steel expanded metal mesh of grades 430 & 304 or grid welded metal mesh. Riblath type meshes should be avoided as the concentration of metal at overlaps can induce cracking in the render due to too greater degree of expansion and contraction of the metal within the render. Suitable overlaps must be provided where necessary of approximately 100mm.

The expanded meshes usually are supplied in sheets 2400mm x 1200mm or rolls 5m x 1m wide and are expanded to a "diamond" pattern of approximately 50mm x 19mm.

The meshes are supplied by specified weights.

All system terminations should be provided with metal accessories or stop trims to include movement joints and points where the system abuts dissimilar materials such as windows, doors, electrical boxes, water pipes or other fixtures.

Avoid damaging the mesh, trowels and floats can indent or distort the mesh if not carefully used.

Never cut the reinforcement to compensate for wrinkles or folds.

Pre-cut reinforcement mesh to workable lengths using a metal cutter. Allow for overlaps of 100mm. Only cut sufficient mesh to cover a suitable working area to enable full application in a reasonable working period.

Install manufacturers pre-formed corner mesh or beads using the basecoat material or mechanical fixings. Use a builders level to check vertical and horizontal alignments.

Place the pre-cut mesh on the wall and fix with mechanical fixings through the insulation board into the substrate, at the prescribed rate. Avoid wrinkles and gaps wherever possible.

Glass or Mineral Fibre Mesh All mineral fibre meshes must be installed correctly in order that they reinforce the basecoat to provide, water resistance, minimise cracking and improve impact resistance.

Glass or mineral fibre mesh, is supplied in rolls 50m x 1m wide. Suitable overlaps must be provided where necessary of approximately 100mm.

The meshes are supplied by specified weights.

All system terminations should be provided with metal accessories or stop trims to include movement joints and points where the system abuts dissimilar materials such as windows, doors, electrical boxes, water pipes or other fixtures.

Pre-cut reinforcement mesh to workable lengths using a cutter. Allow for overlaps of 100mm. Only cut sufficient mesh to cover a suitable working area to enable full application in a reasonable working period.

Install manufacturers pre-formed corner mesh or beads using the basecoat material or mechanical fixings. Use a builders level to check vertical and horizontal alignments.

Place the pre-cut mesh on the wall bed into the basecoat and fix with mechanical fixings through the insulation board into the substrate at the prescribed rate. Avoid damaging the mesh, trowels and floats can indent, cut or distort the mesh if not carefully used. Never cut the reinforcement to compensate for wrinkles or folds these must be smoothed-out with basecoat render.

Application of Basecoats

Base Coat First coat, where required, should be applied to walls in the normal fashion with hawk and trowel.

a. 8-10mm and to manufacturer's specification for metal systems.

b. 4-5mm and to manufacturer's specification for mineral fibre systems.

232

Skim over with the basecoat material any exposed meshed areas to completely cover the mesh, lightly scratch horizontally ready to receive the finish coat.

It is important to take special care to straighten with a darby/straight edge to ensure the next coat is applied to uniform level. Allow 24 hours curing time before further application, unless advised otherwise.

Acrylic basecoats are supplied pre- mixed in buckets and are used directly from the bucket onto the insulation. Some proprietary brands do require the addition of portland cement to complete the mix.

These acrylic systems are applied generally in accordance with the requirements for mineral fibre reinforced systems. See individual manufacturers for systems using these materials.

Finish Coats

Finish coats are applied in accordance with manufacturers or suppliers instructions, details of finishes are explained further in this manual.

5.9   *Simulated Brick-Effect Renders*

Featured System by Kilwaughter Chemical Co Limited

Pre-Contract It is recommended that a test panel (ideally 2 square metres) be produced for inspection, and the client, architect, main contractor, clerk of works etc. should let work commence when completely satisfied with workmanship, texture quality, colour, depth of coat. This should not be part of the main job and must be prepared well in advance to allow time for curing.

Renderers must become familiar with the product water requirement, work-ability characteristics, setting and hardening time which vary according to background, temperature and humidity. This is important as these factors will possibly effect colour uniformity.

Preparation, Design and Application The "working day" areas of application will be dictated by the skill of the operatives, weather conditions, orientation of the works together with any architectural requirements.

Where possible, application on individual wall surfaces should be completed in one operation. Where this is not practical, day work joints to be agreed with the architect.

Simulated brick render is normally applied in two layers over Base Coat. i.e. base coat, mortar coat and face coat, either by trowel or projection render machine.

The face coat is applied prior to the curing of the basecoat, producing a monolithic render.

Texturing   To texture the surface finish, use the appropriate tool, stiff brush, comb, sponge, spatula or other implement as required, but do not over-trowel, polish surface or apply water during set. The skill of the operator will determine the finish.

Cutting and Marking-out   Measure and mark-out the coursing of the bricks to a constant gauge of 65mm with 10mm mortar joints. Measure horizontally and mark-out brick lengths at 210mm with 10mm mortar joints. Differing sized bricks, soldiers and cut bricks are measured accordingly.

After the face layer has been shaded and textured, and following initial stiffening of the applied materials, the face layer is cut through into the basecoat layer, using the appropriate cutting tool. This reproduces recessed mortar coursing of the brickwork or stonework; spirit levels, templates and straight edges should be used for guiding this operation.

Experience will dictate the best time for the operation to take place; too soon and the spirit levels and other guides will mark and spoil the surface, the cutter will rag and tear the material; too late and it becomes difficult and then impossible to cut. At the correct time a clean cut is easily achieved.

The brick style or bond is chosen prior to commencing cutting. Always mark-out the courses first. Always work from the edges and calculate where any headers or "cut" brick can be best sited. Check vertical "perps" at every stage to ensure vertical alignment.

NOTE :- It is not easy to repair brick patterns once cut so advanced planning of brick bonds is essential to avoid irregular brick bonds.

Brushing   Following further stiffening of the materials and using a soft bristled brush, lightly brush and remove any face materials left by the cutting out process, taking care not to bruise surface skin of face materials.

Colour Uniformity 50mm dia samples are provided on request for colour indication only.

Render materials are manufactured from natural products, and slight shade variations may occur.

All areas must be scraped at the same stage of readiness, as early scraping will result in darker shades, and late scraping in lighter shades. A uniform approach is essential to achieve an even finish.

Lime Bloom As the cement sets, lime is produced, and this may come to the surface causing some areas to exhibit an opaque whitish layer.

Render is most susceptible to lime bloom in the early stages of setting, and therefore needs to be properly protected from the weather and any other sources of water.

Lime bloom is a temporary phenomenon, and does not affect the durability or strength of the render.

Do not render in cold damp weather conditions.

Do not permit down-pipes, sills, copings and scaffold boards to throw water on to the setting render.

Do not allow washings from quoins, sills etc. to run on to the setting render.

Design Considerations Suitably designed overhangs and flashings should be provided to prevent water leakage on to the render. At ground level it is recommended that the rendering should not bridge the damp proof course to form a capillary path for rising damp.

Sills and Copings should project from the face of the wall with an ample drip groove for their full length (38mm recommended),, to ensure that dripping water is kept clear of the render.

Plan ahead to avoid discontinuity in anyone area or walling which could lead to unsightly day joints in the rendering.

Render expansion joints are used to keep areas of render to reasonable dimensions to avoid excessive movements.

Application Before any rendering begins, is it is essential to ensure that the scaffolding provides suitable access to the whole of the working face.

If conditions dictate, when rendering on to a base coat, water spray may be used to damp down walls prior to applying the final coat. This will control suction between the two coats.

Renders have a working temperature range of 5°C- 35°C.

During hot weather it is recommended that work is started on the shady side of the building and continued round following the sun, thus avoiding rapid drying causing accelerated curing and the possibility of render cracking.

In cold weather, if frost is forecast, work should stop in time to allow the material to set sufficiently to prevent frost damage. Drying conditions will vary accordingly to wind, temperature and humidity.

Protection from rain and frost should be provided for the first 48 hours after application.

First Coat -Standard Base Coat   First coat, where required, should be applied to walls in the normal fashion with hawk and trowel.

Thickness must be minimum 8-10mm and to system/architect's specification.

It is important to take special care to straighten with a darby/straight edge to ensure the next coat is applied to uniform level.

Allow 24 hours curing time before further application, unless advised otherwise.

Second Coat -Brick Mortar Layer   The coloured mortar layer is applied to the prepared basecoat, 6 -8mm thick, using hawk and trowel or projection render machine to line and level.

Do not over-trowel or polish surface.

Final Coat -Brick Face Layer   After the mortar layer has started to stiffen, the coloured face layer is applied. It must not be applied if the mortar layer has been allowed to dry fully and/or set.

The face layer is applied to the mortar layer following its initial stiffening, using hawk and trowel, or projection render machine. It is applied to an average thickness of 3-5mm and immediately textured. It can also be shaded using dry powder dyes.

The brick layout is cut into the uncured face coat to the design and pattern required as previously described. Brush down only when dry.

Note :- Cutting mistakes into the face coat are not easy to remedy, so care must be taken to plan ahead, the brick pattern, to avoid possible repairs or re-facing the top-coat.

Simulated Stone-Effect Renders　　As above for the application of renders, textures and cutting methods.

However, the stone pattening shall be determined by the required style and effects traditionally provided by the real stone effect to be replicated.

5.10 Insulated Render

Featured System by Wall-Reform

Insulating render enables those buildings which, due to location, design and practical detailing prevents the application of a thick cladding to insulate to the highest standards. Insulated render, by its very nature, does not provide the best thermal resistance but will provide a thermal resistance far greater than render alone.

These renders will improve the thermal standard of the property, improve it's visual appearance and improve the durability to acceptable levels of condition.   The system is applied using traditional methods of application and can be finished with a wide variety of traditional finishes.

Up to 50mm thickness can be applied in two passes. Standard rendering beads to be as required.

The technical data is as follows:-

Apparent density of set mortar	= 204.6kgm
Water vapour resistance	= 1.04mng-1
Thermal conductivity	= 0.055w/mk
Fire Class 0	Tested to BS 476-6/7

The application is of an insulating basecoat, a pre-mixed dry mineral based mortar in accordance with BBA Certificate no 02/3951.   The product consists of standardised mineral binders, graded aggregates and proven additives.

Standard finishes can be applied as follows :-

Dry Dash Aggregates

Tyrolean Finish

Scratch Plaster

Roughcast

5.11 Application of Wet Finishes

The application of cementitious render finishes can be applied to all standard cementitious basecoats, either reinforced with metal or mineral fibre meshes. The acrylic type renders usually are only finished with a pre-mixed, bucket supplied, texture coating.

If in any doubt contact the particular manufacturer or supplier for further clarification.

Traditional   Sand/Cement   Render   Traditional sand/cement mixes produce strong relatively impermeable coatings and applied onto an impermeable backing such as insulation slabs, have a relatively high drying shrinkage. Their use should be restricted to the application onto expanded metal meshes and to areas of a maximum of 5000mm in any direction. A water-proof additive is recommended to the basecoat.

The materials in traditional renderings should be proportioned as to produce a reasonably dense, void-free structure and to have suitable work ability. Strength and durability are provided by the cement so the differing layers should reduce in strength as they are applied. This will reduce the possibility of cracking.

Although lime does not contribute to the strength of renders in the early stages it does provide increased work ability, making it more cohesive and easier to apply. The lime also enables the rendering to retain water against the suction of any background, helps to promote the hydration of cement preventing rapid drying out thus reducing shrinkage.

The usual traditional mixes used onto externally applied insulations are as follows:-

Reinforced layer - basecoat 10mm   1 : 3-4
                              (inc plasticizer & water proofer)

Straightening coat - 6-8mm   1 : 5-6
                              (inc plasticizer)

Top decorative coat -6-8mm   1 : 7
                              (inc plasticizer) (Pigmented)

The top decorative coat can be finished by dash, scratch, Tyrolean or smooth treatments.

Polymer Renders   Over recent years, specialist render suppliers have developed polymer enriched pre-blended and pre-bagged pigmented renders to provide the industry with suitable waterproof renders allowing specialist insulation contractors the freedom from site mixing. They can apply renders of consistent quality that will perform the function of protecting external insulation for a period of some 25-30 years.   These renders are essentially cementitious based to traditional type mixes but with added polymers to improve waterproofness, flexural strength, colour and work ability.

The application is entirely traditional with conventional working practices and tools, however, the open time for dashing is controlled. The scratch renders are designed for scraping and the colours are very consistent due to computer controlled manufacturing methods.

Traditional polymer renders and high polymer base coats are supplied as a complete designed package with full inter-dependent compatibility.

Aggregate Dash Finish   The dash receiver, usually a polymer modified pre-bagged mix, is applied to the base coat to an approximate thickness of 6-8mm, straightened and trowelled smooth.

Coloured aggregates, specially graded and washed for the dry-dash process, are then thrown, in the traditional manner, onto the wet dash-receiver render.   The process involves scooping the aggregate from a "bucket" and throwing it, in a way to maximise the spread, onto the wet render obtaining an even spread.

The aggregate thus applied should be lightly tamped into the dash-receiver with the face of a wooden float, ensuring a good bond.

The size of aggregate should be compatible with the dash receiver thickness i.e. the greater the size of stone, the thicker is the requirement for the receiving render.

Scratch-Plaster Finish   Scratch Plaster contains a courser aggregate to provide, when scratched, an open textured surface.

This type of finish requires the render to be scraped or scratched sometime after the application is applied, the timing largely dependent on the prevailing weather conditions at the time of application.

The specialist render is applied onto the base coat approximately 11-12mm thickness, straightened and trowelled smooth.

When the correct conditions apply, i.e. when the render can be scratched without clogging the scratching tool, the render is scratched in circular motions to remove the top 1-2mm of render and obtain an even textured finish.   Complete areas only should be scratched as scraping at differing drying conditions will produce differing colours and textures.

After scratching the render should be well brushed with a suitable stiff brush to remove surplus material. If small areas are missed in the scratching process, these will appear as a lighter colour and will affect the overall appearance.

Roughcast Render Finish A tradition method, mainly applied in Scotland, this finish is available in a variety of colours and aggregate size, polymer enriched and pre-bagged. The roughcast render is applied onto a prepared coloured render coat applied approximately 6-8mm thickness, this layer being applied onto the standard prepared base coat.

The roughcast render is a mix of coloured render and large aggregate to give a smooth surfaced "nobbly" finish, the aggregate providing the size of texture. Applied by throwing onto the coloured base fully wet roughcast render, thoroughly mixed to an even constituency, to fully cover the coloured base layer. The thickness of the finish roughcast layer will be determined by the aggregate size. This finish can be over-painted fairly easily after fully curing.

Spray Render or Tyrolean Finish Spray or Tyrolean Render is a polymer modified, pre-blended and pre-bagged mix fully designed to be spray applied either with powered mechanical equipment or by hand spraying tools.

The process, as with for roughcast, is applied onto a coloured polymer rendered layer applied 6-8mm thickness and trowelled smooth.

The spray application is to apply an open textured finish, created by 2-3 passes with the spraying tools, to build-up a textured finish, approximately 3-4mm thickness, to that required by the Architect or client.

Usually available in a wide variety of colours, this is a traditional finish widely found throughout the UK.

Smooth Render Finishes   Utilising the polymer renders available for either dashing or for scraping, these renders can be smooth finished, although care is needed to sponge the finish in order to remove trowel marks.

Textured Coat Finish   The textured finish coat is the surface coating which can be used to provide colour and texture and to the external wall insulation system. Most finish coats are acrylic or silicon based, containing a variety of aggregates of differing grades, are factory formulated and supplied in buckets ready to use. Thorough mixing may however be required to avoid settling or separation of the constituents in the bucket.

The specified colour and desired texture must be consistent throughout whole elevations.   Hand tools such as rollers have traditionally been used for textured finish coat application, but spray application is becoming increasingly popular.   Both methods seek to produce an exterior surface that is firmly bonded to the base coat and consistent with the specified architectural finish for the project.

A continuous application of finish coat material is necessary to achieve an overall uniform appearance. If more than one finish is to be used on the building, their respective locations should be clearly defined to all applicators. Logical starting and stopping positions must be established prior to commencing.

Finish coats should never be installed in temperatures below 5degC or onto surfaces that are frozen.   Applications should be applied when temperatures are rising from 5degC.

The finish coat should never be applied during inclement weather such as rain, fog, sleet or snow. During the curing period, installed finish coats must be protected temporarily from sudden weather changes.

Acrylic and silicone finish coatings depend on water loss by evaporation rather than a chemical reaction to dry and cure. They are affected by temperature and relative humidity, wind and solar orientation as well as application thickness. Under ideal conditions the finish coat should by dry in 24-48 hours but rain, fog, or moist weather can extend drying time considerably, warm dry conditions will shorten drying times.

Application of the finish coat should be performed on dry clean surfaces and away from direct sunlight whenever possible, the preferred method is to precede the sun so that the finish is applied in shade and will cure easily after application. The finish should never be applied to wet or dirty surfaces or be allowed to be wetted immediately after application.

If there is any likelihood of inclement weather, protection of the finish should be implemented quickly to avoid contamination. Conversely, protection may be required for strong sunlight as rapid curing may be detrimental to the finish.

Paint Finishes   Paint finishes are generally applied onto smooth render, achieved in a variety of ways previously described. The surface of the render shall be clean and free from any deleterious matter, voids or holes shall be filled and any cracks sealed with a suitable flexible and paintable filler. Any other defects in the surface shall be filled, smoothed-out and sealed as required.

A primer/sealer coat is applied to seal the surface and when dry completed with one or two coats of finishing paint. All to be in accordance with the manufacturers instructions.

5.12 *Application of Dry Finishes*

Tile Hanging   Traditional tile hanging can be used effectively with external wall insulation systems as the depth required by traditional tile hanging methods is compatible with modern external wall insulation compositions.

Vertical support timbers are fixed to the substrate at 400mm centres.

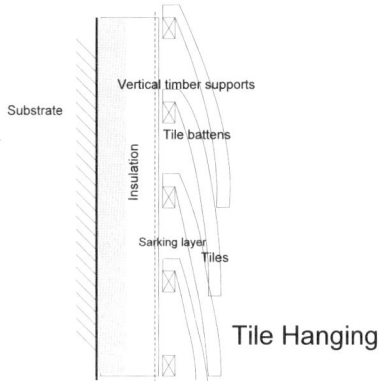

Tile Hanging

Insulation is fixed between the supporting timbers.

Tiling battens are fixed to the vertical supports onto a layer of sarking.

Typical Tile-hanging

Tiles are hung and fixed to the tiling battens.

Boarding   Horizontal boarding can be installed onto timber supports complete with an insulation layer.

Vertical support timbers are fixed to the substrate at 400mm centres.

Insulation is fixed between the supporting timbers.

Shiplap boards are fixed to the vertical supports over a layer of sarking.

If vertical boards are required then the main supporting

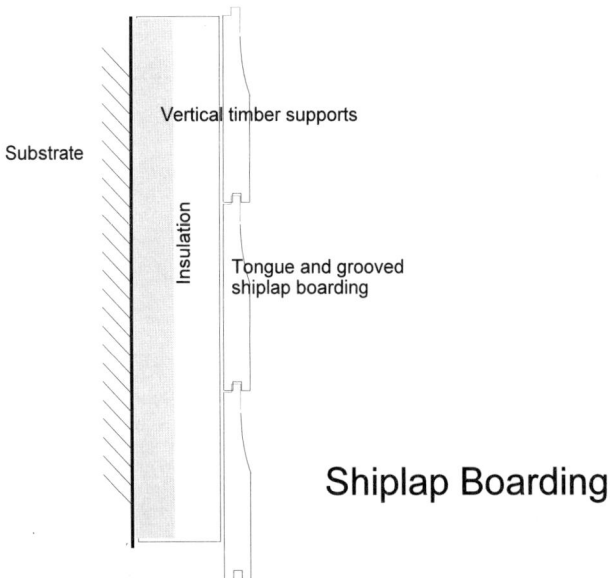

Substrate

Vertical timber supports

Insulation

Tongue and grooved shiplap boarding

Shiplap Boarding

timbers are installed horizontally.

Ship-lap boarding can be treated soft-wood, painted to requirements or cedar for the natural look.

248

5.13 Brick Slips

Brick slip systems are based on various types namely :-

1.   Profiled insulation

2.   Reinforcing mesh carrier onto insulation

3.   Profiled laminated insulation

4.   Resin moulded thin slips - 5-7mm thickness

Once the carrying system is in place the principal of fixing the slips is similar to all types of system.

1.   Select the brick bond to be used.

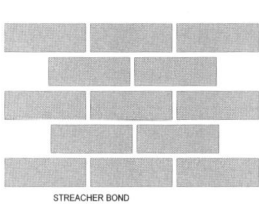

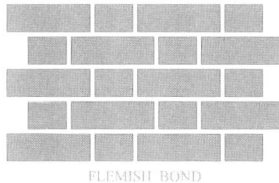

STREACHER BOND   FLEMISH BOND   AMERICAN BOND

Other bonds can be used as specified or required.

2.   Set-out the brick slip system of courses and patterns.

Create horizontal DATUM lines round the complete building, using a water-level or other such equipment and mark with a marking pen so that these lines are easily visible. The datum lines should meet when continued round the building, if they do not then check why not before proceeding. Only continue when true datum lines are properly established.

Measure from these datum lines vertically to the base trim to establish the levelness of the trim. Measure likewise to any other horizontal trim or feature to establish variations in dimensions. It is

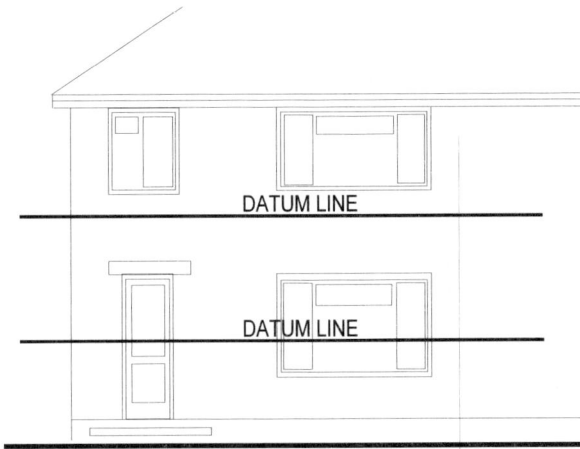

Typical setting-out plan

only possible to proceed with the layout of the brickwork when variations in dimensions are established. Brick layouts are designed utilising soldier courses as and when necessary to correct dimensional irregularities.

3.　　　　　Working from external corners bed the corner slips or "pistols" onto the corners alternating the slips course per course.

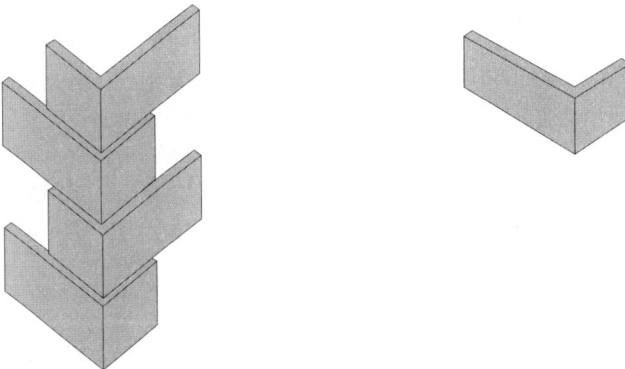

4.   "Soldier" Courses over windows and doors are used to correct the courses if at variance with the adjacent courses. Corner "pistols" can be used with the "short leg" of the slip extending under the window head soffit.

Soldier Course

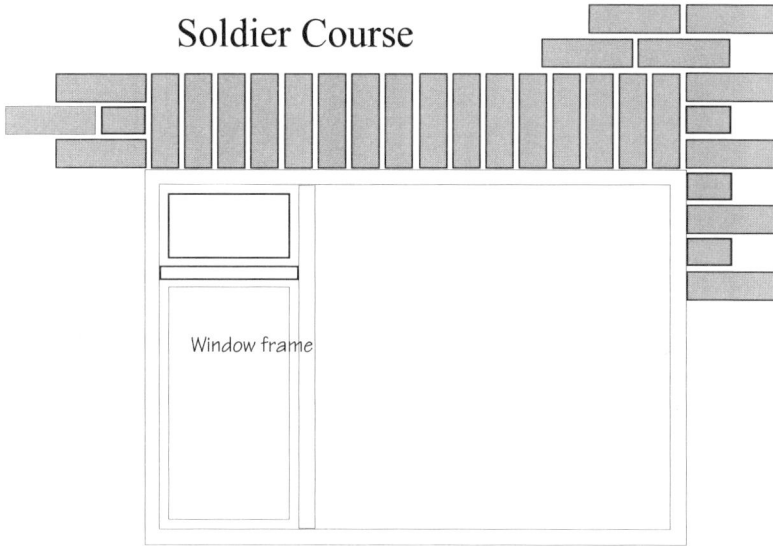

Window frame

"Soldier Course" over window

5.   Brick slip quoins can be used to create good visual effects.

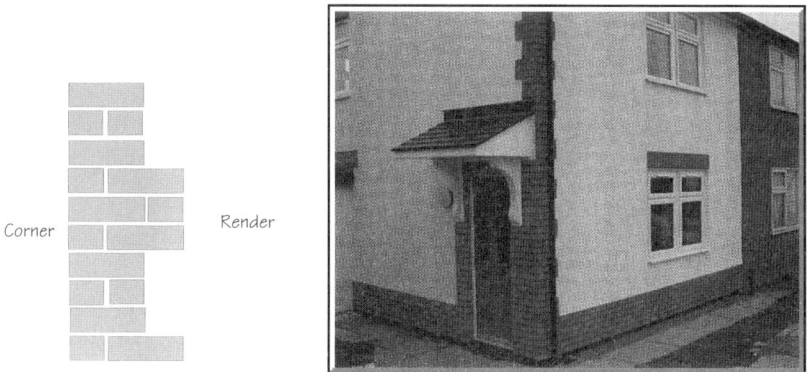

Corner   Render

Brick slips & Soldier Course

Wetherby Systems

6.   Pointing to be installed in accordance with pointing mortar manufacturers instructions.

Pointing brick slips

Note :- The resin moulded slips are adhered to the backing coat which is smoothed and shaped to form the joint.

Brick Slip Joints

Various pointing finishes can be achieved to suite requirements namely :-

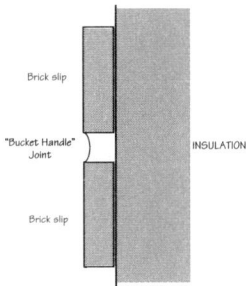

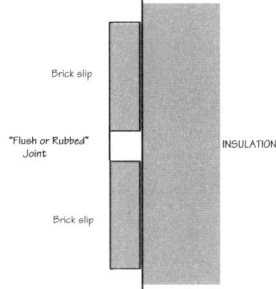

"Bucket Handle" joint *(most common)*   Flush Joint

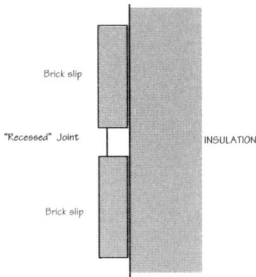

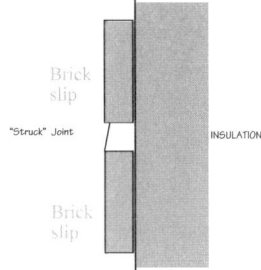

Recessed Joint *(dependent on slip thickness)*   Struck Joint

Brick Slip Project - Wetherby Systems

Special Brick Slip Features

Raised joints & Quoins

Special features such as raised brick joints and protruding corner quoins can be replicated as shown.

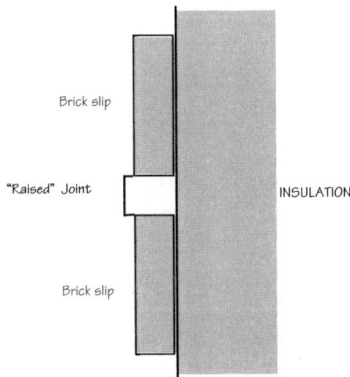

5.14 Sealants

Sealant shall be gun applied to all window reveals, door reveals and other edges of the system where there is no metal profile, before the application of the finishing coat of mortar.

Surplus sealant shall be removed to give a neat surface finish.

All types of sealants to be selected in accordance with manufacturers recommendations and installed in accordance with manufacturers instructions.

5.15 Protection of the Works

The avoidance of many problems on the work site can be alleviated with the installing contractor or employing contractor providing all of the necessary protection of the elements of new works whether installed or partially installed.

Protection will take various forms namely:-

1. Correct positioning of free-standing scaffold. Scaffolding must be erected to allow the wall claddings to be installed and free of internal obstruction from bracing poles. Generally the gap between scaffold poles and the walls to be insulated is 300mm. Any variation from this must be agreed between the installing contractor and safety officer.

Other forms of scaffold, such as mast-climbers, should be installed to provide similar workable facilities.

2.   The use of debris netting, particularly on high rise projects. Most high-rise projects, i.e. those of three storeys or higher, require that debris netting be used to cover all of the external faces of the scaffold.

3.   Low-tack tape installed on all horizontal surfaces (removable). This procedure is recommended to be implemented on powder-coated surfaces such as cills and copings to protect the finish and colour from dropped materials such as wet mortars or thrown aggregate finishes.

4.   Crash barriers at low-level. Barriers are recommended to protect systems in areas subject to severe impact and abrasion. These barriers should take the form of securely fastened metal rails, erected at suitable heights to provide adequate protection to the system from vehicles such as forked trucks etc..

5.   General site protection to exclude vandals. Sites in areas that are noted for vandalism should be well protected with the provision of vandal-proof fencing.

6.   Suitable storage areas for materials. Sites in areas of risk by weather or theft should be adequately protected with the provision of covered and lockable storage facilities.

7.   Temperature checks both high and low. During periods of extreme temperatures both high and low temperatures should be recorded and reported to the contract supervisor as both high and low temperatures can be deemed "inclement" under contractual terms.

8.   Suitable precautions to prevent any detrimental effect of high and low temperatures should allowed for in any contractual arrangements.

5.16 *Common Problems*

Due to the combined use of differing materials, the whole range of design options and the variety of applications, various installation problems can occur. To complicate matters further the core insulant is covered by the finishes so any in-depth examination of apparent problems will require a degree of destructive exploration.   Non-destructive examination is limited in scope to localised visual defects..

Surveying visual problems by a competent surveyor is relatively straightforward using original design techniques and data, conclusions may be more difficult when this data is not available. Contractual considerations are also important when investigating problems such as guarantees, both for the complete system and individual material guarantees which may have to be called-in.

Rendering Defects   Defects in rendering can be either easily visible or disguised from view. The more easily viewed defects, such as cracking, can be remedied as required with either replacement or over-coating, the more difficult defects such as degradation of renders, are those not easily viewed and may only be remedied with total replacement.

Mastic seal defects   Mastic seals can be dislodged by accidental damage, extremes of weather or poor installation. Attention is urgently required under these circumstances as the seals are protecting the system from water ingress and any failure to repair may cause more serious damage not visible elsewhere in the system over long periods of time.

Trim or bead defects   Rusting of galvanised beads close to the surface of rendered finishes is a common defect, whilst difficult to replace economically, beads that rust will only create ever

257

increasing problems with full replacement usually the only option available.

Powder-coatings may de-laminate from the trim surface causing un-sightly finish and painting may be the only economical method of repair.

Bulging coatings   Texture coatings incorrectly applied, or applied onto an unsuitable surface, may "bubble" causing ultimate failure of the finish.

Efflorescence or Lime Bloom   In the construction industry the term "efflorescence" is often used to describe white deposits or stains on finishes. Efflorescence is a general term and as such, covers a number of different deposits varying significantly in chemical composition and method of formation.

To avoid confusion, it should be appreciated that :

1   There exist many kinds of efflorescence, several of which have little in common except the fact that they produce white colourations.

2   The processes which give rise to efflorescence on cementitious renders are generally not identical with those responsible for efflorescence on clay bricks.

3   Colour differences on Render surfaces are not always due to efflorescence. For example colour variations on new structures may be due to late/early scraping, too rapid hydration of render etc. On older structures, unsightly stains are often the result of the contrast in colour between washed and unwashed areas; areas washed by rainwater remain clean while unwashed regions pick up dirt.

On cement-based products, efflorescence normally takes one of three forms :

A   Lime bloom

B   Weeping *(Not normally in renders)*

C   Crystallization of soluble salts

Of these, the most common is lime bloom, lime weeping is mainly confined to concrete forms.

Lime Bloom - This is an occasional phenomenon particularly noticeable on the surfaces of coloured render products made with Portland cement. It is a white deposit which is apparent either as white patches or as an over-all lightening in colour. The latter effect is sometimes mistakenly interpreted as colour fading or being washed out.

The cause of lime bloom lies in the chemical composition of Portland cement. When water is added to cement, a series of chemical reactions takes place which result in setting and hardening. One product of these reactions is "lime" in the form of calcium hydroxide. Calcium hydroxide is slightly soluble in water and, under certain conditions, it can migrate through the damp render to the surface and there react with carbon dioxide from the atmosphere to produce a surface deposit of calcium carbonate crystals. This surface deposit is similar to a very thin coat of whitewash and gives rise to the white patches or lightening of colour as mentioned previously. The surface deposit is normally extremely thin and this thinness is demonstrated by the fact that, when the render is wetted, the film of water on the surface usually makes the deposit transparent and the efflorescence seemingly disappears.

The occurrence of lime bloom on render tends to be spasmodic and unpredictable. Nonetheless an important factor is the weather. Lime bloom forms most readily when render becomes wet and damp for several days, and this is reflected in the fact that it occurs most frequently during the winter months. In particular, an extended period of rain, snow which lies for some time, and damp foggy days are conditions most likely to bring on a severe outbreak.

Lime bloom is not visible on damp render and so only becomes apparent with the onset of drying weather. Thus the drying weather does not necessarily produce the lime bloom; it may only make visible a deposit which had already formed, but could not be seen because the render was damp.

Render is normally only liable to lime bloom early in its life. In general render which has been in service for a year without being affected can be considered immune. Lime bloom is a temporary effect, and given time, usually disappears of its own accord. It is purely superficial and does not affect the durability or strength of the render .

Crystallization of Soluble Salts This type of efflorescence, which corresponds to that normally observed on clay brickwork, is relatively rare on render. It usually takes the form of a fluffy deposit and tends to occur either on render which has been made with sea-water or on retaining walls.

Unlike lime bloom, the deposit is not calcium carbonate but instead consists of soluble salts not normally present in render. These soluble salts can originate from contaminants present in the original render mix, eg sodium chloride introduced by using sea-water as mixing water. Alternatively they may have migrated in the render from external sources, eg substrates, ground-water in contact with walls or foundations. They are drawn to the surface and deposited at places where water evaporates from the render.

Where soluble salts are present as contaminants in the render mix, efflorescence often becomes apparent as the render initially dries out. Where the salts originate from an external source such as ground-water, the rate of build-up of efflorescence depends upon many factors including the permeability of the concrete and the concentration of salts in the ground-water. In some cases, the efflorescence periodically disappears and reappears with changes in weather conditions.

Deposits of soluble salts on the surface do not harm render. However they may be an indication of potential problems, e.g.

If the salts are sulphate, there may be a possibility of sulphate attack of the render. If the salts are chlorides, their presence could, in certain circumstances, bring about corrosion of steel in reinforced concrete or render carriers.

If the render is weak and porous, salts may also be crystallizing under the surface, a condition which could give rise to disruptive stresses and spalling.

Prevention and Removal For over a hundred years the problem of lime bloom on render has been recognized and considerable effort has gone into finding ways of preventing it. Nonetheless there still exists no generally acceptable, guaranteed, preventative measure.

Carbonation need not necessarily result in a change of colour. If the calcium carbonate produced by carbonation is deposited below the surface, it is not visible and does not affect the appearance of the render. Only when it is deposited on the surface does it hide the under-lying colour and give rise to the white colouration known as "lime bloom" or "efflorescence".

Prevention of the lime bloom on render requires measures which discourage a calcium carbonate deposit from forming on the surface, and there are several different approaches which may achieve this.

Control of Curing Render which has been well cured is less liable to lime bloom. Unfortunately, however, conditions conducive to good curing can also be ideal conditions for formation of lime bloom. The aim should be to keep the humidity high, but not so high that lime bloom occurs.

With on-site methods of curing, problems also occur. Spraying with water or the use of damp hessian are not to be recommended where lime bloom would be a problem. Even the use of polythene sheeting for curing can often result in a patchy lime bloom unless it is possible to arrange for the polythene to be slightly separated from, and not in direct contact with the render surfaces. Spray-on curing membranes are preferable, but these can also affect the surface appearance.

Surface Treatments Various surface treatments are available which will reduce the tendency for lime bloom to form. Firstly, there are water-repellant materials such as silicones. These are applied to the surface of render and encourage water to run off rather than remain on the surface. By preventing water from remaining on the surface, the conditions conducive to lime bloom are avoided. Secondly, polymer coatings on the surface of render can be very effective in preventing lime bloom. Such coatings seal the surface and prevent the lime bloom from coming out. Acrylic-based coatings are often used and may be either clear or pigmented. These are generally satisfactory, but occasionally can result in an unacceptable glazed appearance to the render. Also, clear coatings are less durable than pigmented ones. When coloured coatings are used, it has to be accepted that they will eventually weather off, and so their colour must match closely that of the underlying render if unacceptable patchiness is not to result when they are lost from the surface. It is preferable to apply the coatings by spray rather than brush, since brush-applied coatings tend to show up the brush marks as they weather off. By the time the coating is lost, the render should normally be mature enough to be no longer susceptible to lime bloom.

N.B. Some coatings are unsuitable for use on render due to its alkalinity, and some water- based paints can make efflorescence worse.

Finally, treatment with acid is often used to remove lime bloom. It has been found that acid treatment can also act as a preventative measure if applied before bloom has the chance to form. Surface treatment of concrete products may interfere with the bond between the products and mortar and this can have an undesirable side- effect. This is not, however, a problem with acid treatment.

Avoiding Conditions Conducive to Lime Bloom
New render is particularly susceptible to lime bloom and so there can be justification for taking special precautions early in its life. With site conditions, practical considerations usually mean that protecting the render from rain and dew for a few days is difficult. Ideally, protection should be offered. However, the protection should not be in direct contact with the render so that it can trap a film of water between it and the render, as this could result in white patches where the protective sheet touches the render.

Unless adequate provision can be made for ventilation, polythene sheeting of coloured render is not recommended, since condensation inside the film can result in unsightly patches of lime bloom.

Removal of Lime Bloom-Efflorescence   Lime bloom is usually a transient phenomenon and can be expected to disappear with time. The major factor influencing its duration is the environment to which the render is exposed. Where the render is fully exposed to the weather, rainwater (which is slightly acidic) dissolves the deposit and the lime bloom typically disappears in about a year. In more sheltered locations, removal by natural means may take considerably longer .

If immediate removal is required, this can be achieved by washing with diluted acid. This is a relatively simple operation, but care should be taken on two accounts. Firstly, acids can be hazardous and appropriate safety precautions must be taken. Secondly, acid attacks render and over-application to a render surface can result in acid etching, which will alter the texture and appearance.

Generally, an 8% solution of commercial grade of hydrochloric acid is used. The acid concentration can be adjusted to suit individual circumstances; a less concentrated solution will require more applications to remove lime bloom but will be less likely to result in an acid etched appearance.

Before the acid is applied, the surface should be dampened with water to kill initial suction. This is best achieved by the use of a steam cleaner used cold with a pressurised spray. This prevents the acid from being sucked into the render before it has a chance to react with the surface deposit.

The acid is applied by spray using a "killaspray" type garden spray and a typical application rate is one litre of acid 5-10 square metres. Following application of acid, the surface of the render is immediately power washed at approx 70 deg C. Often one wash with acid is sufficient, but in more stubborn cases the treatment is repeated as necessary until the lime bloom disappears.

When carrying out acid washing, always start with a trial on an inconspicuous area. Operatives should wear protective clothing, at the very least rubber gloves and goggles. Precautions should be taken to prevent acid from coming into contact with metals and other materials which may be adversely affected.

Acid is neutralised within seconds of coming into contact with render; consequently when acid washing is used on render products there is no risk of acid burns to users of such products. The attack on render by acid, even in the case of severe over- application, is limited to a thin surface layer and there need be no cause for concern that acid washing will affect properties of the render other than the surface appearance. Whilst there can be no guarantee, experience suggests that lime bloom is unlikely to re-occur following its removal with acid.

If acid washing is used to carry out remedial action to renders, care must be taken if there is any possibility of the acid coming into contact with aluminium or PVC. These materials can be found in the construction of windows, cills, copings, nosings and others. Acid will certainly attack aluminium and may also affect PVC, dependent on the grade encountered.

Prevention and removal of deposits of soluble salts As mentioned previously, salts crystallizing on the surface of render may originate either from impurities present in the water used to mix the render or from ground-water in contact with render. Efflorescence resulting from contaminations present in the render mix is often a result of using sea-water as mixing water. The use of sea-water should be avoided in all situations.

Ground-water does not migrate very easily through good quality render and soluble salts from ground- waters do not often crystallize on render surfaces. In situations where precautions are considered necessary, a bitumen (or similar) damp-proof membrane should be used to separate the render from the ground-water.

These deposits are often soft and fluffy and in many cases can be removed by using a dry bristle brush. Should this fail, the next treatment to try is to combine brushing and washing with water. should this also fail to remove the deposit, the surface should be washed with acid as described above. In all cases, trials on an inconspicuous area should be carried out to determine the most effective treatment.

5.17 *Repair & Maintenance*

Building owners are often not aware that a problem exists until long after extensive damage has occurred, external wall insulation systems which comprise a variety of differing materials can accumulate significant quantities of moisture before any problem becomes apparent. Seals of mastic, whilst apparently still present and appearing sound, may not function correctly if poorly installed or have suffered attack from severe weather conditions or frost.

Most repair and maintenance procedures will require the use of established external wall insulation components and wherever possible all materials should be supplied by the original system supplier. If this is not possible, then the closest possible compatible alternative materials should be used.

The repair and maintenance of external wall insulation systems is the result of two essential problems, accidental damage and system failure. Accidental damage is usually damage caused by the hard impact of a moving object causing the system to be altered sufficiently as to not perform its intended use, or to be visually impaired so as to be unacceptable.

Whether the system is damaged or has failed, the first report by a client or architect should signify action by associated professionals in order to determine the exact nature or extent of such damage or failure in order that the necessary remedial action can be taken. Suitable examination, which may require some destruction of the surrounding wall area, will be required in order that a report be prepared so as facilitate the production of bills of quantities and/or estimates to enable repairs to be completed. Material or system guarantees together with contractual terms and conditions with system designs may need to be examined to assess liabilities.

When inspecting reported problems it is important to understand the complex nature of certain installations and the fact that one problem, apparent in one location can lead to other problems in other locations. This is particularly evident when examining water penetration due to failed seals, water may have penetrated into unexplained associated areas causing damage which as yet has not become apparent. Once the originating problem has been ascertained, the proposed repair plan can be drawn-up with further investigation in-mind to help eliminate or reduce possible future hidden problems. It may be that periodic inspections are advised at reasonable intervals in the future to monitor the system for future performance.

Inspections Before proceeding with any particular course of action, it is essential to report and analyse the defect. Hammer test areas and look for the cause of problems in order that a repair may be assessed. It may be that the initial inspection will call for further detailed examination, this can be referred to as a "destructive" examination as part removal of the finishes or subsequent layers may be required to ascertain the exact nature of the defect or damage. Internal water damage caused by leaking pipes can cause hidden damage to insulation layers and it is likely that a "destructive" inspection may be necessary under these or similar circumstances.

Non-destructive testing in the form of small-scale probes to determine system thicknesses or small sample taking for analysis in a laboratory may achieve significant results without the client suffering wholesale visual destruction of his wall finishes.

Inspections of the following to be carried-out:-

Insulation - Inspect for damage, moisture content, or degradation - replace as necessary.

Renders - Inspect for any or all of the following using the hammer test procedure together with visual inspections and positive probing:-

1.   Damage - however caused - persistent damage may be removable, incidental damage may be more difficult.

2.   Thicknesses - to design expectations - additional layers may be needed to correct this problem dependent on circumstances.

3   Degradation - of render surfaces by weathering or other causes - renders may have to be replaced as necessary.

   ### *Usual causes*

   Poor quality render

   Poor workmanship

4   Moisture content (insulation)- can be excessive due to external problems which may have to be remedied to avoid persistent un-wanted moisture saturation.

   ### *Usual causes*

   Failed sealants

5   Heavy spalling - of all render coats exposing sub-strate - may require over-coating. Look for extensive random crazing, moss growth, poor keying, bad sub-strate and inconsistent or poor thicknesses.

   ### *Usual causes*

   Frost attack due to poor weather resistance

   Water penetration behind render

   Poor sub-strate

   Chemical and air borne attack

6   Light spalling - of finished renders shedding finishes

   ### *Usual causes*

   Frost attack due to poor weather resistance

   Water penetration behind render, examine plumbing and detailing

   Poor sub-strate

   Chemical and air borne attack

7   De-lamination - of multi-coated renders and finishes leaving poor keying

   ### *Usual causes*

   Lack of key - first coat not properly scratched

   Too rapid drying of render coats

   Little or no water-proofer in base-coat

   Top-coat stronger than base-coat

8   Heavy cracking - more than 1mm wide, often around openings sometimes   reflecting brick or block patterning.

   ### *Usual causes*

   Settlement or structural cracking

   Shrinkage due to cracking of brick or block substrate

   Differing base materials

   Note :- heavy cracking is not normally confined to render, it is usually a defect in the sub-strate.

9   Light cracking - often around doors and windows - examine for any moss, growth in cracks, defect most obvious when the surface is drying out.

Usual causes

Base coat not fully shrunk before top coat is applied

Incorrect sand used - too high a shrinkage

Too much fine material

Too rapid drying in hot weather

Top coat too strong

Poor installation of reinforcing mesh

10   Balding - resulting in loss of surface texture or dashing aggregates

Look for

Shape of aggregate

Severe exposure of building - wind

Age of render

Usual causes

Old age - weather

Failure of aggregate to bond to top-coat

Failed top-coat too weak to hold aggregate

11   Friable renders - easily rubbed away by hand over sizable areas

Look for

Loss of top-coat thickness

Loss of aggregate - examine floor adjacent for shed material

Surface easily rubbed-away

Old age

Lack of waterproofer in render coats

Too much air entrainment

Coat frosted after application

12 Substrate spalling - surface renders detaching complete with backings

Usual causes

Water penetration through cracks

Poor/inadequate preparation and keying to substrate

13 Wrinkled surface - dry dash slumping - fluid effect

Usual causes

Mix too soft when applied

Scaffold too close

Dash thrown too soon

14 Lift marks - shadowy bands corresponding with scaffold levels or other obstacles more common on dry-dash

Usual causes

Rain hitting scaffold boards causing mud splashes onto finished surfaces/inadequate protection of finished work

Scaffold too close

Scaffold left in place too long after work completed

15 Scud or whip marks - fan shaped patterns usually found on roughcast

Usual causes

Applicator's imperfect technique

Scaffold too close on application

16   Bleaching - applies to coloured mortars, areas of white coming through

Usual causes

Rain damage before initial set has taken place/inadequate protection of working area and completed work.

17   Colour variations - applied to coloured mortars/renders

Usual causes

Renders mixed incorrectly

Renders applied in varying weather conditions

18   Lime Bloom - applied to all coloured cementitious mortars and is evident with white shadowy "blotches" discolouring the finish

Usual causes

Temperature and moisture inconsistent with the correct curing particularly before the initial-set has taken place.

Rust stains - resulting from galvanised products deteriorating within the render

Usual causes

1.   Galvanised finish being damaged before renders applied

2.   Un-protected products being installed

3.   Stray nails in surface

4.   High iron content of aggregate

5.   General contamination of render during mixing

6.   Contamination from surface fixings, avoided by careful detailing.

Coatings   Inspect for damage, de-lamination, colour degradation, colour variation

Trims   Inspect for damage, poor fixing or sealing

Beads   Inspect for damage, poor fixing/detailing

Brick Slips   Inspect for poor adhesion, poor pointing, water penetration

Mastic Seals   Inspect for poor adhesion, undersized mastics or missing seals, incorrect sealants installed.

Repair methods should follow the original system specification as far as it is possible in order to maintain the original design characteristics of the completed system.

Where the system cannot be followed materials should be used to match the existing as near as it is possible.

5.18 Repairing the System

The very nature of the assembly of numerous components to make-up the External Wall Insulation System means that there are layers of differing materials, all requiring their own repair methods. Access to each layer is essential in order that re-assembly can be assured, each layer must be repaired independently and correctly.

To achieve the access to the deepest material, each layer must be removed in ever reducing sizes but sufficiently large enough to ensure each layer has sufficient area to fix or adhere adequately to the lower layer.

The principal of layering the system to expose the various elements may require large areas of the system to be exposed, this is dependent on the size and nature of the project and its access ability.

Repair methods - Insulation Boards Remove any damaged insulation board, preferably to a natural straight joint, and replace with a board of similar manufacture and type. Where boards have been adhesive fixed, the old adhesive may have to be removed to assist in the installation of the new board which should also be adhesively fixed.

Where the insulation board has been mechanically fixed, new fixings, similar to the originals are to be used into newly drilled holes.

Gaps in insulation boards should be avoided, any gaps created will need to be filled with injected foam, cleaned off flush after curing.

Repair methods - Metal Reinforcements   Remove the damaged render to the surrounding areas exposing the metal reinforcement to approximately 100mm beyond the area of damage, examine the insulation board and repair as necessary.   This is to expose the metal reinforcement to undamaged surfaces to enable the metal to be cut in reasonable panels to enable repairs to be implemented.   Cut the existing metal in straight lines if possible to allow the backing render to be cut away a minimum of 75mm beyond the metal leaving the metal free in open air.

Apply new basecoat material to the area exposed, including below the original metal reinforcement newly exposed, trowel as flat as possible, lightly scratch the exposed areas.   Cut new metal reinforcement to allow for an overlap of approximately 100mm. Fix with mechanical fasteners to the minimum rate of 8/9 per m². Apply topcoat render as required.

Repair methods - M/F Reinforcement   Remove the render around the damaged area up-to a suitable natural "stop" location to help reduce the patched look of the repair, examine the insulation board and repair as necessary. Remove total render thickness together with the damaged mesh and repair or patch the base coat as necessary.   Renew the mineral fibre mesh overlapping by a minimum of 100mm, use fasteners as necessary.   Trowel the basecoat over the mesh and finish with a light horizontal scratched finish ready to receive the finish coat.   Apply the finish coat as the original specification.

Repair of renders can produce a "patch" or "patched" effect, this should be eliminated as far as is practical by only repairing the rendered area up to natural stops such as corners or trims or beads.

Repair Methods - Brick Slips   Remove and discard the damaged brick slips together with any support system and examine the insulation board for any damage, replace/repair as necessary. Install as needed any brick slip carrier/support system in accordance with the original specification. Replace slips and point the joints.

Use only similar brick slips, if possible, to those originally installed, if this is not possible it may be that a designed panel using alternative slips may have to be implemented to make the repairs visually acceptable.

Repair Methods - Texture Coatings Existing textured coatings are not easily removed without the removal of base coats, damaged areas may require the repair of insulation boards which can be replaced as previous. The texture coating may require to be applied to complete wall surfaces in order that colours and textures are retained. The original system supplier may need to be consulted in respect of colour and texture matching, grain size may be critical in obtaining a reasonable texture match.

Repair Methods - Mastic Seals Mastic seals should be physically checked periodically to establish that the mastic is still in place and that the performance, expected of the mastic, to weather-proof the system is maintained. Should any damage occur then the mastic seal can be replaced with a similar application. Only good quality proprietary mastics, recommended for this application should be used and applied in accordance with the mastic supplier's instructions.

Repair Methods - Trims and Accessories Trims and accessories include all exposed trims used to assist in weather-proofing the system, including, overcills, undercills, copings and edge/verge/top capping trims. These trims are usually powder-coated aluminium. Welded components are included as required.

To repair cills or copings, the external wall insulation system may have to be removed to facilitate the replacement fixing straps which will be required to be fixed direct to the supporting wall or structure.

Impact Damage   Impact damage is usually at low level and in relatively small areas and usually caused by vandalism or impact by vehicles.

If the damage is localised and the insulation layer is intact, patching the render may be sufficient. If the insulation layer is damaged then replacement is essential.

A system of equal performance is required to be used for repairs, however, the make-up of the system can be varied to suite circumstances, particularly if a more substantial system of repair is deemed sensible. Finishes then become important to try to blend new and old to reduce detrimental visual impacts.

Patching Renders   Render patching can be problematic particularly if materials are difficult to obtain of a like nature to the adjoining areas. In these cases it may be prudent to consider rendering complete areas to avoid visual clashes of the use of incompatible materials.

5.19 Attachment of External Components

External fittings such as :-

Satellite Dishes

Clothes lines

Gates & Fencing, road signs etc..

Light fittings etc..

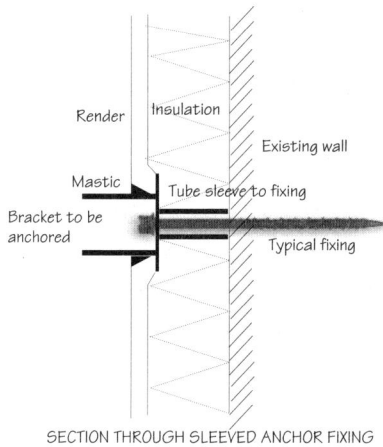

SECTION THROUGH SLEEVED ANCHOR FIXING

Are from time to time required to be fixed to the face of the external cladding. Strong anchors may be required but these can depress or compress the system causing faults and water ingress.

Any anchors so used must be "sleeved" to prevent this depression or compression in accordance with the drawing.

Some external attachments such as light fittings may be more suitably fixed to a timber patress in order that waterproof electrical connections can be made within the depth of the patress and adequate seals applied.

PART SIX - Application

6.1 Tendering & Design

Tenders submitted by prospective applicators or contractors are to be calculated on the basis of the project conditions and in strict accordance with the specifications, bills of quantities (if supplied) drawings and samples supplied by the Client, Architect or Surveyor.

Where BBA/BRE certification is required for the external wall insulation system, particular regard must be given to compliance with the terms and conditions contained within the relevant certification.

The principals of tendering and design should take into account all of the following:-

1. Thermal performance of the existing wall and/or the whole building envelope (S.A.P. Method)

2. Preparation works

3. Accessibility of the works

4. Existing service pipework and ducting requirements

5. Thermal performance of the completed insulated wall

6. Durability

7. Thickness of the overall system

8. Finishes & Colours

9. Costings of specified and unspecified materials

10. Time of year when project is to be carried-out and any requisite protection to the ongoing and completed works.

11.   BBA/BRE Certification Terms and Conditions

12.   Size of project - economy of scale

13.   Requirements of Main Contractor (if any)

14.   Phasing - both installation and hand-over to completion

15.   Guarantees

The existing walls are to be surveyed to check their existing thermal performances together with any associated problems such as water penetration, ability to fix the system, architectural features, features needed to be replicated, (if any), difficult abutments, protrusions and miscellaneous fixed but removable/replaceable items such as satellite dishes and road signs should also to be included.

6.2   Selecting a System

The external wall insulation system to be selected is dependent on several factors relating to the project requirements. These are as follows :-

a.   Determination of the thermal performance necessary to improve the project to the requisite standards.

b.   Determination of the durability requirement in consideration of the project's location and exposure, particularly to weather conditions or vandalism.

c.   Determination of the necessary finish required to comply with any planning, environmental or architectural requirement.

Certification After consideration of the above, certification by the BBA/BRE must be considered to give credibility to any project as to its system reliability and overall performance.

Various guarantee schemes are available, particularly through INCA, provided the system chosen is suitably certified.

Fixing the System   Adhesives are generally used when a lightweight system is specified and where the substrate is suitable to adhere-to, being:-

1.   Clean and free from any deleterious material, hammer test wall surfaces for any de-lamination of substrate

2.   Fairly flat and free from any severe undulations and high/low spots

3.   Readily accessible to allow suitable trowelled application

4.   Where the building or building use does not require any fire-resistence consideration, particularly when using a combustible insulation such as polystyrene.

Mechanical fixings, either plastic, plastic and metal or all metal fixings will require a "pull-out" test to determine the size, number and type of fixing to be used. The following will assist in deciding the type of fixing to be used:-

1.   Clean and free from any deleterious material

2.   Dub-out wall surfaces where required

3.   Carry-out "pull-out" test on a selection of fixings - choose the best results for application.

Note :- Consult the Code of Practice CP3 for Wind-Loadings to all projects to assist in the selection of the correct fixing or method of fixing.

Selecting the Finish & Colours Attention to detail is required in the selection of finishes paying due regard to the local Planning requirements.

Additional considerations may be necessary if the project involves a Conservation Area, Listed Building, Area of Outstanding Natural Beauty or National Parks Authority.

Finishes are selected from the following :-

Aggregate Dashing & Coloured Backings

Scraped Render or Scratch Plaster

Smooth Surface Renders

Painted Smooth Renders

Tyrolean Renders

Textured Coatings

Dry Finishes -   *Timber Ship-lap*

                            Terra Cotta Tiles

                            Tile Hanging

                            Panel Finishes

Brick Slips

Simulated Brick using Renders

Simulated Stone using Renders

6.3   Selecting the Contractor

The contractor to be selected is preferably one which is a member of the Insulated Render & Cladding Association, the trade association dedicated to oversee the industry. Contractors who are members of this association have proved their capability, workmanship and financial standing to the association.

6.4   Site Drawings

Site drawings consisting of elevations, detailing and finishing schedules should be prepared prior to the instigation of any project, they should include some or all of the following:-

1.   Location & Positioning of Beads and Trims

2.   Location of movement joints

3.   Colour and texture schedules/elevations

4.   Flashing details

5.   Window & Door Cill details

6.   Architectural Features

     Quoins

     Bandings - Raised

     Bandings - Recessed

     Ashlar cuts

7.   Coping details

8.   Special details

     Window Pod Details

     Abutments

     Extensions

9.   Attachments - such as :-

     Gates & fences

     Clothes lines

     Satellite dishes

6.5 *Manufacturing Drawings - Special Components*

Special components such as window cills and copings will require site measurement and custom manufacture to enable the correct designs and manufacturing standards to be achieved.

Drawings should include the following :-

1. Scaled details of all adjacent abutments/openings

2. Detailed design of components to be installed

3. Dimensions sufficient for manufacture and installation

4. Material specification

5. Material thicknesses

6. Material coatings

7. Finishing colour

8. Quantities

9. Site location plan of products - where necessary

10. Installation instructions

11. Fixing instructions

12. Delivery Schedules

6.6   *Site samples*

A pre-installation site sample is a valuable means of obtaining approval from the client as to colour acceptability and overall quality assurance. The site sample is constructed to full project specification, the larger the sample the better and more detailed the approval can be obtained, however, samples of the basic system measuring 1.5x1.5m are normally acceptable.

The site sample should include all of the basic project details including the following :-

1.   Sealants

2.   End terminations

3.   Outside and inside corners

4.   Joints

5.   Cappings

6.   Cills

7.   Finishes

8.   Abutment of differing finishes

9.   Colours and textures

10.   Beads, trims & nosings

The sample should remain on site for the duration of the contract, it may be possible, on the right project, for this sample to be incorporated within the project contract to keep costs to a minimum.

The cost of such a sample is usually incorporated within the project cost at tender stage.

6.7   Pre-Contract

Before the contract commences, pre-contract site meetings should be held between the following :-

1.   Main Contractor.

2.   Specialist Insulation Sub-Contractor.

3.   Client or Architect or Project Administrator.

to determine the following:-

1.   Programmes - To include   Delivery Schedules

Project Phasing

Hand-over Schedules

& targets

2.   Site conditions.

3.   Storage of materials.

4.   Welfare facilities for workmen.

5.   Architectural features.

6.   Re-establishment of any fixtures and fittings.

7.   Power & water supplies.

It is important to discuss the junction with adjacent finishes and building operations to ensure continuity.

It is also important to discuss protection of the working area and the finished work together with the responsibility of who is to provide such protection. The costs, of protection, if any, should also be agreed, together with the supply responsibilities at an early stage of the work.

6.8 Material scheduling & Ordering

The Specialist Insulation Sub-Contractor shall be conversant with all of his specialist materials and prepare his requirement schedules well in advance of the project commencement. This is to ensure that any bespoke items requiring special manufacture be completed correctly and on time, e.g. aluminium cills and copings. These specialist materials are usually supplied in following manner:-

Insulation boards	by packs (vary by type)
Fixings	by box - usually in 200s
Trims	2500mm lengths (typical)
Beads	3000mm lengths (typical)
Cills & Copings	Usually to surveyed requirements
Renders	by bags 25kg weight (typical)
Coatings	by drums 15kg weight (typical)
Aggregates	by bags 25kg weight (typical)

All material should be inspected on delivery, particular attention should be given to the following:-

All materials should be checked to ensure the correct quantities are delivered and that these items are checked against the material orders to establish any shortages or incorrect deliveries.

All bagged material should be checked for broken bags, batch numbers and shelf-life dates.

All trims & beads to be checked for damage to straightness, size, type and colour.

All insulation boards to be checked for visual damage, correct type, grade and thicknesses.

All boxes to be checked for conformity with orders, types, codes and sizes.

All bucket materials to be checked for damage, code numbers, colours and shelf-life.

Technical information to the supply of renders, insulation boards and coating materials should include the manufacturers name, address, plant address, batch number, date of manufacture and shelf-life where appropriate.

Certificates of Conformity in respect of specialist materials such as stainless steel should be provided as required.

6.9   *Storage of Materials*

All materials should be stored in dry conditions, preferable in covered storage facilities protected from the weather as well as accidental damage incurred by other trades.

All cementitious materials are damaged by damp/wet conditions, coating materials are damaged by low temperatures and all materials are suspect to accidental damage by others.

Pallets carrying insulation materials should be stored off the ground, if stored outside, and the shrink-wrapping maintained as long as possible so as to offer full protection to exposed edges.

6.10 *Site Inspections*

Site inspections are usually carried out by the system promoter or sponsor to verify that the installation is being installed correctly and to specification to ensure that both his and the clients interests are being protected. The following stages should be checked for conformity to specification:-

1.   Adaption/removal of existing services/ducting/flue extensions etc..

2.   Sub-strate Preparation works

     to include, keying, scarifying, removal of old paint surfaces etc..

3.   Trims fixed directly to substrate.

4.   Insulation Boarding

5.   Fixing pattern & Pull-out tests

6.   Render beading

7.   Basecoat render and reinforcing meshes including detail meshes

8.   Finishes

6.11 *Site Reports*

Site reports may be requested by the client and listed here are those reports usually requested:-

1.   Substrate examination/testing report.

2.   Pull-out test report on fixings.

3.   Sample testing of renders for performance verification.

4.   Colour testing and matching of coating finishes.

5.   Weather reports to include high/low temperatures during the contract period.

6.12 Monitoring Job Conditions

Project weather conditions are usually monitored in respect of temperature and rainfall. The temperature will affect the curing of renders and when below 5deg C may adversely affect the curing of water-based coatings.

Work will generally require to cease when the temperature is falling and reaches 5degC, it may resume when the temperature commences to rise and reaches 5degC. These are as a guide only as local conditions and orientations may have to be taken into account. To monitor these conditions temperature readings are taken daily at a mutually agreed time of the day to suit local circumstances. Records from local weather centres may also be referred-to.

Rainfall will also affect the performance of renders being applied on a project. Whilst still wet, any water-based render or coating will suffer a "wash-off" requiring replacement or reinstatement dependant on circumstances. The action of this effect may also be dependent on exposure.

Recording rainfall and temperature level is essential if the programme falls behind due to weather conditions, particularly if claims are made for contractual delays.

Materials used in waterproofing and finishing external wall insulation systems are by nature tough and durable, so spillages may be difficult to remove if cured. Monitoring the care and attention taken to ensure the protection of adjoining areas of work from such spillages will enable the final clean-down on completion to be easier and more satisfactory.

Protection of work can be achieved with the application of suitable coverings in the appropriate places, cementitious materials will cure very hard and coating materials can easily stain so adequate protection can improve the overall quality of the final finish.

6.13 *Certification of Completion*

A Certificate of Practical Completion is usually issued by the Architect or Contract Supervisor when fully satisfied as to the completion of the project installation that complies with the issued specification. Detailed lists of outstanding small items still to be completed may be part of this certificate.

6.14 *Guarantees*

Guarantees or warranties should be issued on completion of the works or immediately after any period of maintenance. These guarantees are for a normal period of 10 or 20 years and can include a scheme operated by INCA.

Guarantees or warranties should include the following:-

1. The term of the guarantee or warrantee will be (?) years as specified.

2. The system will provide the thermal resistance specified.

3. The exposure of the systems is suitable for (?) exposed conditions.

4. The system will be suitable for exposure to (?) impacts.

5. The system will remain secured to the wall.

6. The finish will require maintenance after (?) years.

7. The seals will require maintenance after (?) years.

8. Any powder-coated finish will be guaranteed in

accordance with the manufacturers instructions.

9.   Any proprietary finish supplied will be guaranteed in accordance with the manufacturers instructions.

6.15 Owner's Manual

It is useful for the Building Owner or Occupier to be informed of the details of the external wall insulation system and what he can or cannot do to it. The manual should include the following :-

1.   Details of the Main Contractor, Specialist Installing Contractor complete with contacts.

2.   Details of Certifications (BBA or BRE)

3.   Details of any guarantees and the method of making a claim.

4.   Instructions on casual maintenance

5.   Details of how to fix external items such as :-

>   Satellite Dishes
>
>   Fencing
>
>   Gates
>
>   Signs
>
>   Any other common item

6.   Instructions on what not to do to the insulation system. (May invalidate any guarantee)

6.16 Access, Tools & Safety

Tools for application of Systems   The skilful use of the correct tools, proper selection and maintenance is required to ensure quality of application and finish. The majority of projects are still installed with the use of hand tools and conventional power tools, however, spraying equipment is increasingly used to apply coating and render finishes.

A typical selection of tools is as follows :-

Hammer Drills

Paddle Mixers

Metal & timber saws

Metal cutters etc..

Scaffold requirements   External Wall Insulation, by its definition is insulating the outside of buildings. This process requires that the wall is free from obstruction to enable the installer to install all of the component parts from preparation to completion of the final finishes.

The requirement of access to the wall is sometimes in conflict with the Health and Safety requirements to keep clearances between scaffold and wall to a minimum. Generally a clearance of 300mm is acceptable to Safety Inspectors and it is just sufficient for the application of wet trades including dry-dashing.

Tubular Scaffold   The installation of external wall insulation usually requires the provision of a suitable free-standing scaffold, situated clear of the building face sufficiently to enable the installation to be constructed but close enough to be safe for the working operatives. It is usual for scaffold poles to be erected 300mm from the face of the substrate, this may vary dependent on circumstances and site conditions, but must be agreed with the specialist insulation contractor.

The site set-up must provide a sturdy structure for the specialist insulation contractor who will need to perform a wide variety of installation tasks. Tubular scaffold should be secured firmly on level ground or suitable provision made to provide a firm bearing on un-even ground. Internal braces should be avoided as far as possible as they can restrict internal usability within the scaffold areas.

All scaffolds must comply with Health and Safety Regulations.

Scissor Lifts Scissor lifts limit an applicator's range of movement to small areas of projects and are usually un-suitable for projects greater than three storeys. Scissor lifts may be more suitable for small areas or repair-work. They can be unstable on uneven ground.

Mast climbers Suitable for high-rise projects, these mobile access platforms allow ease of application to multi-storey projects although their use frequently is restrictive in load capability. Generally workmen are restricted to two at anyone time with weather conditions, orientation, wind and rain also governing their use.

Safety Regulations All current Health and Safety Regulations relating to site access and scaffolds in particular must be complied with.

Particularly relating to :-

1.   Access to the work

2.   Storage of materials

3.   Use of power hand tools

4.   Use of hand tools

5.   Use of mixers

294

6.   Scaffold access

7.   Power to the works

8.   Welfare facilities for workmen

Scaffold Notes

Scaffolding can be a major problem, on site, if not carefully considered in its design and method of erection.

The specific requirements for scaffolding, together with the sequencing of erection and removal is a major area of dispute and difficulties between the external wall insulation installation contractor and the main contractor who often is responsible for its erection.

Additionally there can be conflicts between users of the scaffolding, especially if more than one trade requires to use the same scaffold at anyone time. It is also unlikely that the level of lifts for the external wall insulation contractor will be complient with other users i.e. roofing contractors.

The sequencing of insulation contracts i.e. in groups of say 5 properties, each group to be completed before the next group is started, is likely to be inconsistent with the planned flow of work required for an economical external wall insulation installation contract.

Consideration should be included in any contract for an overlap in the provision of scaffold from one phase to another.

6.17 COSHH Information

(Control of Substances Hazardous to Health)

All site materials should be accompanied by a COSHH report as supplied by the manufacturer or supplier. These reports are to provide all the necessary information to the operative on site together with the site administration personnel, regarding any hazards or methods of application requiring any precautions to avoid the possibility of injury to any site operative.

These reports are to include for any instruction as to the provision of suitable protective clothing including goggles, gloves or foot-ware. Instructions as to any emergency procedures or actions in the event of injury are also stated.

The following typical COSHH notes are indicative of what is expected by the Contractor or site Applicator for the use of the various selected products. These notes are indicative only and may vary from manufacturer to manufacturer and from product to product.

Data References:

Consumer Protection Act 1987

Health and Safety At Work Act 1974

Control Of Substances Hazardous To Health (COSHH) Environmental Protection Act 1991

All materials supplied must display the following information clearly labelled on the packaging:-

Manufacturer

Name and address

Telephone number including any emergency number

Fax number

Email address

Identification of substance or material

Any hazardous reference / emergency action

6.18   Names and Useful Links

Directory of Organisations

British Board of Agrément (BBA)

PO Box 195

Bucknalls Lane

Garston

Watford,   Herts WD2 7NG   Tel :- 01923 665400

BRE Certification Limited (WIMLAS - BRE)

Bucknalls Lane

Garston

Watford,   Herts

WD25 9XX   Tel :- 01923 664603

Insulated Cladding Association (INCA)

PO Box 12

Haslemere

Surrey GU27 3AH   Tel :- 01428 654011

British Standards Institution (BSI)

389 Chiswick High Road
London
W4 4AL   Tel:- 020 899 6900

Insulating Concrete Formwork Association (ICFA)
PO Box 72

Billingshurst

West Sussex

RH14 0FD   Tel :- 07004 5005

Construction Fixings Association

C/O I S T.   Henry Street
Sheffield
S3 7EQ   Tel :- 01142 789143

Directory of Component Suppliers

Trims & Beads, Reinforcement Meshes

Wemico Limited
Matthew Lane
Hoo Farm Ind Est
Worcester Road
Kidderminster
Worcs DY11 7RA   Tel :- 01562 820123

Expanded Metal Co Limited
Expamet
PO Box 52
Longhill Ind Est (North)
Hartlepool
TS25 1PR   Tel :- 01429 866688

BRC Special Products
Carver Road
Ashfields Ind Est
Stafford
ST16 3BP   Tel :- 01875 222288

Renderplas Limited
Stourport Road
Bewdley
Worcs DY12 1BD   Tel :- 01299 400340

VWS Technologie AM BAU
Siemensstrabe 2
D - 72805
Lichtenstein   Tel :- 00497129/695-133

Aggregate Suppliers

Brett Specialized Aggregates
Sturry Quarry
Fordwich Road
Sturry
Canterbury
Kent CT2 0BW   Tel :- 01227 712876

Derbyshire Aggregates Limited
Arbor Low Works
Youlgrave
Nr Bakewell
Derbyshire DE14 1JS   Tel :- 01629 636500

Texture Coatings Manufacturers & Suppliers

Glixton
Carrs Paints Limited
Westminster Works
Alvechurch Road
West Heath
Birmingham B31 3PG   Tel :- 0121 243 1122

A P G Management
Little Paddock
Church Lane
Flamborough
E Yorks   Tel :- 01262 850631

Render Manufacturers & Suppliers

Kilwaughter Chemical Co Ltd
Kilwaughter Works
Larne
N. Ireland   BT40 2TJ   Tel :- 02828 260766

Readymix Drypack Limited
The Sion
Crown Glass Place
Nailsea
Somerset   BS48 1RF   Tel :- 01275 855103

Wallreform Limited
3 Melbourne Close
Marton
Middlesborough
TS7 8NL   Tel :- 01642 272848

Fixings & Fasteners Manufacturers & Suppliers

Hilti

24-26 Great Suffolk Street

London

SE1 0UE   Tel :- 0800 083 0858

Fischer Fixings

Whiteley Road

Wallingford

Oxfordshire OX10 9AT   Tel :-   01491 827 900

Ejot UK Limited

Hurricane Close

Sherburn Enterprise Park

Sherburn-in-Elmet

Leeds LS25 6PB   Tel :- 01977 687040

Insulation Board Suppliers

Kingspan Insulation Limited

Pembridge

Leominster

Hereford HR6 9LA   Tel :- 0870 850 8555

Rockwool

Pencoed

Bridgend

Mid-Glamorgan

CF35 5NY   Tel :- 01656 862621

Sheffield Insulations

Nunnery Drive

Sheffield

S Yorks

S2 1TA   Tel :- 0114 241 3000

Vencil Resil Limited

Infinity House

Anderson Way

Belvedere

Kent DA17 6BG   Tel :- 0208 320 9100

Pittsbugh Corning (UK) Limited

63 Milford Road

Reading

Berkshire RG1 8LG   Tel :- 0118 950 0655

New Build - Permanent Formwork Systems

BecoWallform   www.becowallform.co.uk

Eurozone Building Solutions   www.eurozone.com

Nudura Integrated Building   www.nudura.com

Logix   www.logix.uk.com

Polarwall   www.polarwall.co.uk

Polysteel   www.polysteel.co.uk

Quad-lock   www.quadlock.com

Springvale   www.springvale.com

Styro Build   www.styrobuild.com

Useful Publications

BRE Publications & Digests

Standard "U" Values	Digest 108
Assessment of Wind Loads	Digest 119
Condensation	Digest 110
Heat Losses through Ground Floors	Digest 145
Heat Losses from Dwellings	Digest 190
Energy Consumption in buildings	Digest 191
Cavity Insulation	Digest 236

British Standards Institution

Clay Bricks	BS 3921 1985
Basic data - Control of Condensation	BS 5250 1975
External Rendered Finishes	BS 5262 1976
Fire Characteristics of EPS	BS 6203 1982
Cements	BS 12, BS146, BS4027, BS 5224
Limes	BS 890
Sands	BS 1199
Aggregates	BS 882
Pigments	BS 1014
Plasticizers	BS 4887
Water	BS 3148
Expanded Metal Mesh	BS 1369
Welded Wire Mesh	BS 729
Stainless Steel	BS 1449

Systems currently certified

BBA ref

05/4222	Structherm
03/4022	Structherm
05/4206	Envirowall
04/4136	Wallreform
02/3951	Wallreform
03/4058	Epsicon
00/3766	Ispotherm
99/3564	Thermaloc
98/3548	Dryvit
97/3410	Swisspan
96/3247	Alsecco
96/3243	Structherm
96/3242	BRC
96/3238	Alsecco
95/3213	Testa Teres
95/3132	Sto
95/3090	Powerwall
93/2914	Swisslab
91/2691	Weber
91/2600	Weber
90/2437	Rockshield
87/1800	Permarock

BBA BRITISH BOARD OF AGRÉMENT
CERTIFICATE No 03/4058

Renders Currently Certified

BBA Cert 97/3428

Kilwaughter Chemical Co Limited

(Includes Brick-Rend)

BBA Cert 02/3951

Wallreform Limited

BRE Certified Systems (WIMLAS)

Permarock Phenolic external wall insulation

Technical Approval Certificate 3/5/2002

Permarock External Wall Insulation System

Technical Approval Certificate 3/5/2002

Eurobrick Brick Slip System

Technical Approval Certificate 015/93

6.19 Useful Websites

British Board of Agrément	www.bbacerts.co.uk
BRE Certification (Wimlas)	www.bre.co.uk
Insulated Cladding Association	www.inca-ltd.org.uk
Insulating Concrete Formwork Assoc	www.icfinfo.org.uk
Construction Fixing Association	www.fixingscfa.co.uk
Kilwaughter Chemical Co Ltd	www.k-rend.co.uk
Vencil Resil Limited	www.vencel.co.uk
Derbyshire Aggregates	www.derbyshireaggregates.com
Expamet	www.expamet.co.uk
Ejot UK Ltd	www.ejot.co.uk
Kingspan Insulations	www.kingspaninsulation.com
Hilti	www.hilti.co.uk
Wemico	www.wemico.co.uk
Klimex	www.klimex.nl
SPS BV	www.sps.bv.com
Glixtone Limited	www.glixtone.com
Alsecco (UK) Limited	www.alsecco.co.uk
Alumasc Exterior Building	www.alumasc-exteriors.co.uk
BRC Special Products	www.brc-special-products.co.uk
Envirowall	www.envirowall.co.uk
Dryvit UK Ltd	www.dryvit.co.uk
Permarock Products Ltd	www.permarock.com
Rockwool Ltd	www.rockwool.co.uk
Sto Limited	www.sto.co.uk
Wetherby Building Systems	www.thermaloc.com
Epsicon	www.epsicon.co.uk
Weber Building Solutions	www.weberbuildingsolutions.co.uk
Eurobrick	www.brick-cladding.co.uk
Telling	www.telling.co.uk
Renderplas	www.renderplas.co.uk
Wall-Reform	www.wallreform.co.uk
VWS Technologie AM BAU	www.vws-online.de

NOTES:-

CD ROM Contents
Detailed Drawings. Reports & Checks

Item number

1.	General Arrangement	47	Brick Slips 6 - Pointing Joints
2	Insulation Board Layouts	48	System Junction Detail 1
3.	Insulation Corner Details	49	System Junction Detail 2
4.	Typical Section Details	50	Attachment Details
5.	Rail Fixing Details	51	Quoin Detail
6.	Fixings Layout	52	Coping
7.	Mineral Fibre Mesh Corners	53	Coping Details
8.	Expanded Metal Mesh - Overlaps	54	Ashlar Cuts
9.	Base Trim Detail 1	55	Tile Hanging
10.	Base Trim Detail 2	56	Service Duct 1
11.	Base Insulation Below DPC	57	Service Duct 2
12.	System Corner Detail	58	Service Duct 3
13.	Render Movement Joint	59	Vertical Boarding
14.	Structural Movement Joint 1	60	Drawing - New-Build
15.	Structural Movement Joint 2	61	Insulation to difficult area 1
16.	Abutment Detail 1	61a	Insulation to difficult area 2
17.	Abutment Detail 2		
18.	Top Capping Detail 1		
19.	Top Capping Detail 2		
20.	Verge Detail 1		
21.	Verge Detail 2		
22.	End-Stop Detail 1		
23	End-Stop Detail 2		

The drawings above are intended as a guide only, each project may require variations suitable for that particular project.

REPORT SHEETS

24.	Reveal Detail - Uninsulated	62	Standard Report Sheet 1
25.	Reveal Detail - Insulated		*Pull-Out Test Report*
26.	Reveal Detail - Window Replaced	63	Standard Report Sheet 2
27.	Reveal Detail - Window Repositioned		*Weather Record Sheet*
28.	Window Head - No Drip	64	Standard Report Sheet 3
29.	Window Head - With Drip		*Certificate of Practical Completion*
30	Window Cill - Timber	65	Standard Report Sheet 4
31	Window Cill - Under-cill		*Owner's Manual*
32	Window Cill - Over-cill		
33	Window Cill - Detail 1		

CHECK LISTS

34	Window Cill - Detail 2		
35	Mid-Point Trim Detail	66	System Check List
36	Cill Replacements	67	Design Check List
37	Bandings - External		
38	Bandings - Recessed		**INDEX**
39	Window Pods		
40	Fire Break - Details	68	Index of Sub-Headings
41	Fire Break - Layout		
42	Brick Slips 1 - Bonds		
43	Brick Slips 2 - Soldier Courses		
44	Brick Slips 3 - Quoins		
45	Brick Slips 4 - Corners		
46	Brick Slips 5 - Setting-Out		